COMPUTATIONAL METHODS IN STRUCTURAL AND CONTINUUM MECHANICS

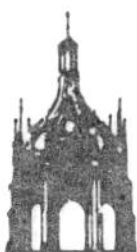

ELLIS HORWOOD SERIES IN ENGINEERING SCIENCE

STRENGTH OF MATERIALS
J. M. ALEXANDER, University College of Swansea.

TECHNOLOGY OF ENGINEERING MANUFACTURE
J. M. ALEXANDER, R. C. BREWER, Imperial College of Science and Technology, University of London, J. R. CROOKALL, Cranfield Institute of Technology.

VIBRATION ANALYSIS AND CONTROL SYSTEM DYNAMICS
CHRISTOPHER BEARDS, Imperial College of Science and Technology, University of London.

COMPUTER AIDED DESIGN AND MANUFACTURE
C. B. BESANT, Imperial College of Science and Technology, University of London.

STRUCTURAL DESIGN AND SAFETY
D. I. BLOCKLEY, University of Bristol.

BASIC LUBRICATION THEORY 3rd Edition
ALASTAIR CAMERON, Imperial College of Science and Technology, University of London.

STRUCTURAL MODELLING AND OPTIMIZATION
D. G. CARMICHAEL, University of Western Australia

ADVANCED MECHANICS OF MATERIALS 2nd Edition
Sir HUGH FORD, F.R.S., Imperial College of Science and Technology, University of London and J. M. ALEXANDER, University College of Swansea.

ELASTICITY AND PLASTICITY IN ENGINEERING
Sir HUGH FORD, F.R.S. and R. T. FENNER, Imperial College of Science and Technology, University of London.

INTRODUCTION TO LOADBEARING BRICKWORK
A. W. HENDRY, B. A. SINHA and S. R. DAVIES, University of Edinburgh

ANALYSIS AND DESIGN OF CONNECTIONS BETWEEN STRUCTURAL JOINTS
M. HOLMES and L. H. MARTIN, University of Aston in Birmingham

TECHNIQUES OF FINITE ELEMENTS
BRUCE M. IRONS, University of Calgary, and S. AHMAD, Bangladesh University of Engineering and Technology, Dacca.

FINITE ELEMENT PRIMER
BRUCE IRONS and N. SHRIVE, University of Calgary

PROBABILITY FOR ENGINEERING DECISIONS: A Bayesian Approach
I. J. JORDAAN, University of Calgary

STRUCTURAL DESIGN OF CABLE-SUSPENDED ROOFS
L. KOLLAR, City Planning Office, Budapest and K. SZABO, Budapest Technical University.

CONTROL OF FLUID POWER, 2nd Edition
D. McCLOY, The Northern Ireland Polytechnic and H. R. MARTIN, University of Waterloo, Ontario, Canada.

TUNNELS: Planning, Design, Construction
T. M. MEGAW and JOHN BARTLETT, Mott, Hay and Anderson, International Consulting Engineers

UNSTEADY FLUID FLOW
R. PARKER, University College, Swansea

DYNAMICS OF MECHANICAL SYSTEMS 2nd Edition
J. M. PRENTIS, University of Cambridge.

ENERGY METHODS IN VIBRATION ANALYSIS
T. H. RICHARDS, University of Aston, Birmingham.

ENERGY METHODS IN STRESS ANALYSIS: With an Introduction to Finite Element Techniques
T. H. RICHARDS, University of Aston, Birmingham.

ROBOTICS AND TELECHIRICS
M. W. THRING, Queen Mary College, University of London

STRESS ANALYSIS OF POLYMERS 2nd Edition
J. G. WILLIAMS, Imperial College of Science and Technology, University of London.

COMPUTATIONAL METHODS IN STRUCTURAL AND CONTINUUM MECHANICS

C. T. F. ROSS, B.Sc. (Hons), Ph.D.
Department of Mechanical Engineering
Portsmouth Polytechnic

ELLIS HORWOOD LIMITED
Publishers · Chichester

Halsted Press: a division of
JOHN WILEY & SONS
New York · Brisbane · Chichester · Toronto

First published in 1982 by

ELLIS HORWOOD LIMITED
Market Cross House, Cooper Street, Chichester, West Sussex, PO19 1EB, England

The publisher's colophon is reproduced from James Gillison's drawing of the ancient Market Cross, Chichester.

Distributors:

Australia, New Zealand, South-east Asia:
Jacaranda-Wiley Ltd., Jacaranda Press,
JOHN WILEY & SONS INC.,
G.P.O. Box 859, Brisbane, Queensland 40001, Australia

Canada:
JOHN WILEY & SONS CANADA LIMITED
22 Worcester Road, Rexdale, Ontario, Canada.

Europe, Africa:
JOHN WILEY & SONS LIMITED
Baffins Lane, Chichester, West Sussex, England.

North and South America and the rest of the world:
Halsted Press: a division of
JOHN WILEY & SONS
605 Third Avenue, New York, N.Y. 10016, U.S.A.

British Library Cataloguing in Publication Data
Ross, C. T. F.
Computational methods in structural and continuum mechanics. –
(Ellis Horwood series in engineering science)
1. Structural engineering – Data processing
2. BASIC (Computer program language)
3. Microcomputers – Programming
I. Title
624.1'028'5425 TA641

Library of Congress Card No. 81-20272 AACR2

ISBN 0-85312-432-9 (Ellis Horwood Limited, Publishers – Library Edn.)
ISBN 0-85312-442-6 (Ellis Horwood Limited, Publishers – Student Edn.)
ISBN 0-470-27329-1 (Halsted Press)

Typeset in Press Roman by Ellis Horwood Ltd.
Printed in the USA by The Maple Vail Book Manufacturing Group, New York.

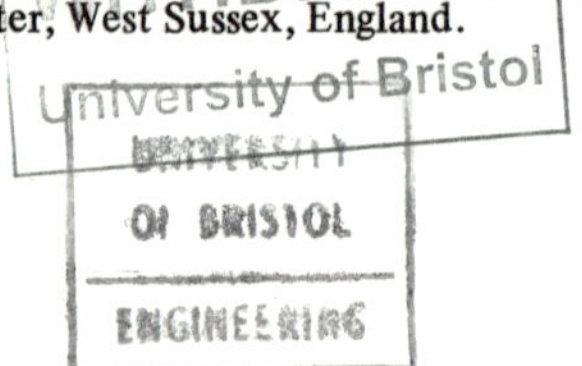

Hi, Thanks for ordering via anybook.com. We hope that you are completely satisfied with every aspect of your purchase but if there are any problems, please give us a chance to put it right by dropping us a line before leaving us any bad feedback. Of course if you are happy with your purchase...please do leave nice feedback!

Please Note: If there are any books from your order that are not in this parcel, they may be delivered in a seperate parcel.

Return address: Anybook.com
2 Outer Circle Business Park, Outer Circle Road,
Lincoln, LN2 4HX
+44 (0) 1522 424108

AP-TPPO1033647-1-0811
800 Avondale Ave
1033647-1-0811
Grandview Heights
OH
43212
USA

CUSTOMS DECLARATION Great Britain	CN22 May be opened officially	
☐ Gift ☐ Documents	☐ Commercial Sample ☒ Other	
Quantity and details description of contents	Weight (kg)	Value
Used books	0.57	£14.00
HS Tariff Number 4901990000	Total Weight 0.57	Total Value £14.00
I, the undersigned, whose name and address are given on the item, certify that the particulars given in this declaration are correct and that this item does not contain any dangerous article or articles by legislation or by postal or customs regulations	09/07/2026	

4926419

3565943

R

Table of Contents

Chapter 13 – Plane stress and plane strain

Chapter 14 – Two dimensional field problems

Appendices

Author's Preface

Fourteen computer programs are presented in CBM PET BASIC, together with descriptions of how to use them.

These programs cover a comprehensive selection of topics in structural mechanics, and a smaller selection of two-dimensional problems in continuum mechanics.

The structural problems include the static and free vibration analysis of two- and three-dimensional structures; the continuum mechanics programs cover plane stress and plane strain, the solution of steady-state problems such as the irrotational flow of a fluid, heat transfer, seepage, the torsion of non-circular sections, and a solution of the two-dimensional Poisson equation with complex boundary conditions.

This text has been prompted partly by the improvements in the number crunching ability of microcomputers, and partly by the rapidly increasing availability of these machines in educational and research establishments and industrial concerns of various sizes.

Indeed, the minimum user RAM required by these machines is only 16 k bytes, thus, the capital outlay on hardware need not be large. Larger machines, however, are preferable as larger problems can be tackled on them, and with the rapidly increasing RAM sizes in microcomputers, it appears that the main limitations will be on precision and speed. Even with a 32 k PET, the static analysis of a continuous beam with 199 nodes took only about 8 minutes.

The book, therefore, fills a gap between the numerous texts that appear in theoretical methods in structural mechanics with little or no software support, and the very sophisticated packages that are available for main frames.

In general, the well known and sophisticated packages for main frames will be superior to the present set of programs for large problems, but for the large numbers of smaller problems, the programs of the present text may be preferred. Furthermore, the skill of using the sophisticated packages is much greater than that required for operating microcomputers, and users are very often advised to attend courses prior to using the sophisticated packages. In any case, many establishments and industrial concerns cannot afford to buy or use sophisticated packages.

This book contains a number of examples, and the user should carefully work through these to thoroughly understand how to use these programs. In addition to this, the user should compare solutions with worked examples from other sources, and take particular care that he/she is fully conversant with sign conventions for loads, etc.

Computer Programs

Although these programs were originally written in microsoft disk on a 16 k Commodore PET, they can be, and are adapted to run on other microcomputer machinery including APPLE. All the programs are now available on a 5¼″ single-side, single-density disk and on cassette, and can be purchased from Ellis Horwood Ltd, Market Cross House, Cooper Street, Chichester, West Sussex PO19 1EB.

Although these programs have been carefully tested, users should satisfy themselves that the criteria of these programs are completely valid and safe when applied to practical problems.

Neither the author nor the publishers can accept any responsibility for any liability whatsoever that might arise through their use.

Acknowledgements

The author would like to thank the Computer Science Department of Portsmouth Polytechnic for their cooperation over many years.

His thanks are extended to his many colleagues for their advice and useful hints. In particular, he would like to thank Mrs Lesley Jenkinson for showing such patience and care in typing the manuscript.

Notation

Unless otherwise stated, the following symbols are adopted:

L	=	length of element or member
A	=	area (usually cross-sectional)
I	=	second moment of area
J	=	torsional constant
E	=	elastic modulus
G	=	rigidity modulus
ρ	=	density
λ	=	eigenvalue
ω	=	circular frequency (radians/s)
n	=	frequency (Hz)
$[\mathbf{K}]$	=	stiffness matrix
$[\mathbf{M}]$	=	mass matrix
$[\mathbf{I}]$	=	identity matrix
$[\ ]$	=	a square or rectangular matrix
$\{\ \}$	=	a column vector

The following shorthand notation is used to denote the input/output of vectors

$$
\begin{array}{l}
\ulcorner\, i = 1(1)N \\
\llcorner\!\rightarrow x_i \; y_i \; z_i
\end{array}
$$

Its meaning is, that the value of i should be increased from one in unit steps to a maximum of N, and for each step, either input or output (as appropriate), the values of x, y and z corresponding to that particular value of i.

To my family

Introduction

Numerical methods in structural mechanics [1-3] first proved of interest shortly after the invention of the high speed digital computer with its own memory. This was a fairly natural process, as numerical methods are very much dependent on digital computers. Indeed, without the aid of digital computers, numerical methods in structural and continuum mechanics would be virtually useless.

During the fifties, there was much debate as to whether the force method was superior to the displacement method or vice-versa, and in 1956, the argument seemed to be settled by Turner *et al.* (4), who produced the finite element method.

The finite element method adopted the matrix displacement method, and extended it to problems such as plates and shells. It represented a very big breakthrough, as up to then, only skeletal structures could be analysed.

Much work was done on the finite element analysis of structures during the sixties [5], and in 1963, Melosh [6] made an important contribution when he linked the finite element method with variational methods.

This prompted many researchers to investigate field problems and in 1971, Zienkiewicz [7] produced a book which covered a large selection of steady state and transient field problems.

Since then, much interest has been shown on energy methods [8] continuum mechanics [9] and more sophisticated topics in finite element methods [10].

For static analysis, the programs assume the matrix relationship:

$$\{\mathbf{q}\} = [\mathbf{K}]\,\{\mathbf{u}\} \tag{i}$$

where,

$\{\mathbf{q}\}$ = a vector of known externally applied loads.
$[\mathbf{K}]$ = a 'constrained' structural stiffness matrix.
$\{\mathbf{u}\}$ = a vector of unknown nodal displacements.

Solution of (i) can be achieved by inverting $[\mathbf{K}]$, and premultiplying it into (ii) to give,

$$\{\mathbf{u}\} = [\mathbf{K}^{-1}]\,\{\mathbf{q}\} \tag{ii}$$

Equation (ii), however, is only suitable for small problems, where the inverse of [K] can be achieved easily, but for large problems, the inverse of [K] can become a major problem.

Fortunately, however, [K] is always symmetrical, and if the nodal numbering is suitably chosen, it may become of banded form, as shown in (iii), where all the non-zero elements appear in a narrow band, symmetrical about the leading diagonal.

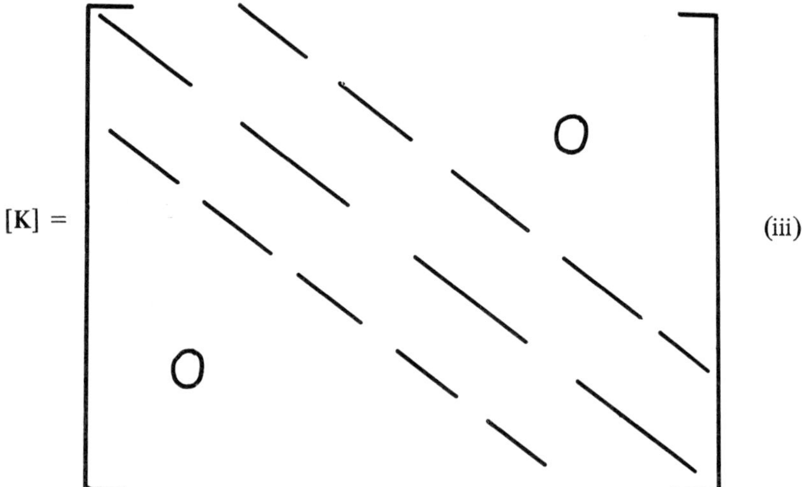

Zienkiewicz [7] and Irons and Ahmad [10], have shown that by considering only the upper half of the band, the solution of (i) can be achieved through Gaussian elimination. It is this method that is adopted in the present text for static problems.

Zienkiewicz has also shown that there are other methods of solving these equations, and so too has Irons and Ahmad.

The flow diagram for static problems is shown in Fig. (i).

If the user prefers to solve the skeletal structures using FORTRAN, he/she may find it very difficult to translate BASIC into FORTRAN. For such cases, the reader is referred to [11].

For vibration problems [12], the matrix equation for free vibration is:

$$|[\mathbf{K}] - \omega^2 [\mathbf{M}]| = 0 \ . \tag{iv}$$

In the present text, as all the problems were constrained or over-constrained, and as the power method was being used, [K] was inverted and premultiplied into (iv), to give the standard eigenvalue equation of (v)

$$|\lambda[\mathbf{I}] - [\mathbf{K}^{-1}][\mathbf{M}]| = 0 \ , \tag{v}$$

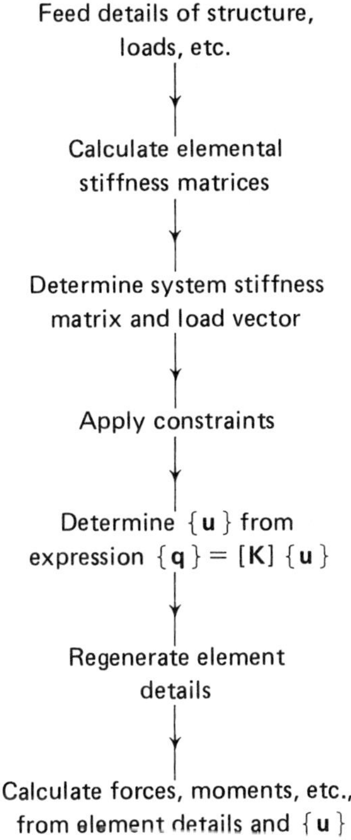

Fig. (i) – Flow diagram for static problems.

where,

$\lambda = 1/\omega^2$

ω = circular frequency

$n = \omega/2\pi$ = natural frequency

[I] = unit matrix.

[K] = system stiffness matrix, with rows and columns corresponding to constraints removed,

[M] = system mass matrix, with rows and columns corresponding to constraints removed.

Thus, the eigenvectors from the programs correspond only to 'free' displacements, the other displacements being zero.

The precision of the eigenvalues was set to 0.1% and this can be changed if desired, but it must be pointed out, that greater precision leads to more computer time, and if the precision factor is too small, it is possible for the precision of the machine to be exceeded. No precision was sought on the eigenvectors.

The flow diagram for free vibrations is shown in Fig. (ii)

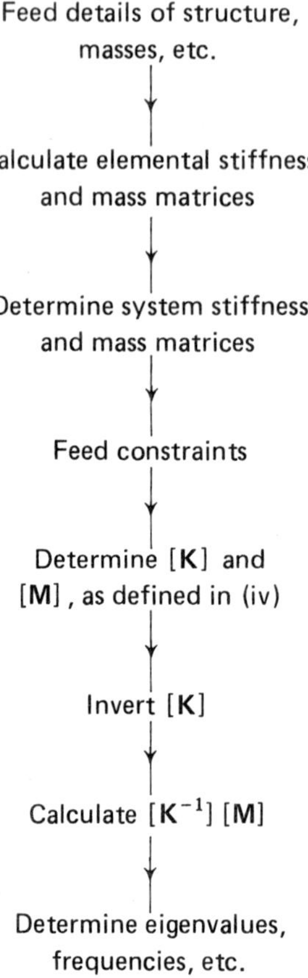

Fig. (ii) – Flow diagram for vibration problems.

Prior to using the programs, users are advised to thoroughly familiarise themselves with the finite element method in structural and continuum mechanics [7,10], together with numerical methods [13] and programming [14].

1

Forces in plane pin-jointed trusses

This program, (Appendix 1), determines the nodal displacements and forces in plane pin-jointed trusses of the type shown in Fig. 1.1. The truss can be statically determinate or statically indeterminate and its members can have different sectional and material properties.

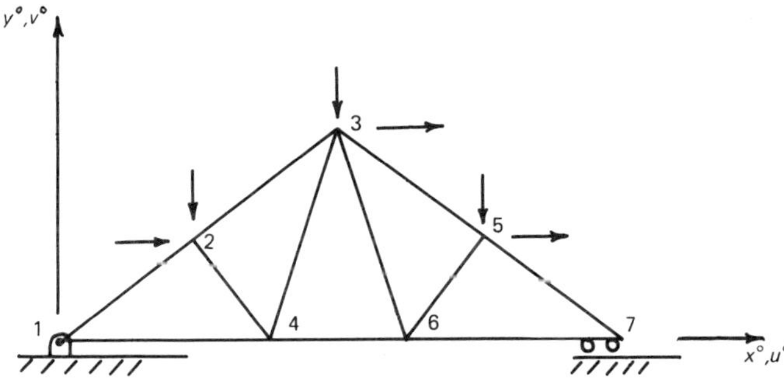

Fig. 1.1.

The nodes are taken at the pin-joints, so that each member is defined by two nodes, as shown in Fig. 1.2.

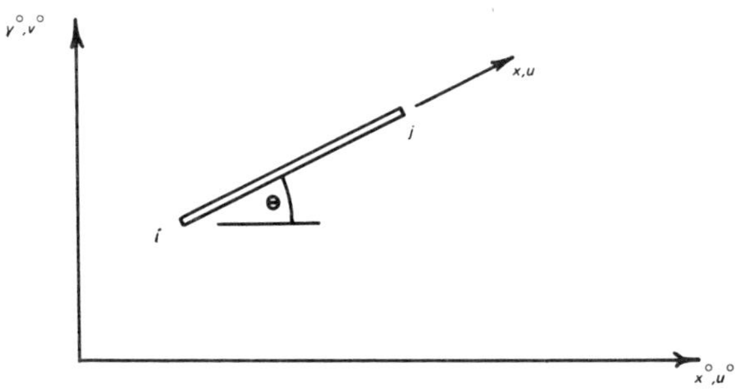

Fig. 1.2.

The members of the truss possess only axial stiffness, and each node has two global degrees of freedom, u° and v°. Thus, for the truss of Fig. 1.1, there are a total of 14 degrees of freedom, and 3 of these are suppressed; both displacements at node 1, (u_1° and v_1°) and one at node 7, (v_7°).

The displacement u_1° corresponds to the displacement position 1 and v_1° corresponds to the displacement position 2. The suppressed displacement v_7° corresponds to the displacement position 14.

These displacement positions are calculated as follows:

$$u_1^\circ \equiv 2 \times 1 - 1 = 1$$

$$v_1^\circ \equiv 2 \times 1 = 2$$

$$v_7^\circ \equiv 2 \times 7 = 14$$

Thus, for the ith node, the displacement position corresponding to

$$u_i^\circ \equiv 2 \times i - 1$$

$$v_i^\circ \equiv 2 \times i$$

1.1 DATA

The data should be fed in as follows:

(1) Number of nodal points or pin-joints = NN

(2) Number of one dimensional structural members = ES

(3) Number of suppressed displacements = NF. (It should be remembered that each node has two global displacements.)

(4) *Positions of suppressed displacements*

┌FOR i = 1 To NF) Input the position of each suppressed displacement
│)
└→Input NS_i) in ascending order.

(5) *Nodal coordinates*

┌FOR i = 1 To NN) Input the x° and y° coordinates of each node.
└→Input x_i°, y_i°)

(6) *Member details*

┌FOR EL = 1 To ES
│ Input i — Input i node for member
│ Input j — Input j node for member
│ Input A — Input cross-sectional area for member
└→Input E — Input elastic modulus for member

(7) *Vector of nodal forces*

┌FOR i = 1 To $N2$
└→Input Q_i

where $N2 = NN \times 2$

1.2 OUTPUT

Vector of nodal displacements

```
┌FOR i = 1 To N2
└→PRINT u_i°, v_i°
```

Nodal forces in each member

```
┌FOR i = 1 To ES
└→PRINT Nodes defining each member and the force in each member
         (tensile +ve)
```

Example 1.1

Determine the forces in the plane pin-jointed truss shown in Fig. 1.3. All members are of constant E, and all members are of constant A, except for member 2-3, which has a cross-sectional area of $2A$.

As only the forces are required, it will be convenient to assume that $A = E = 1$. If, however, the nodal displacements are required, then it will be necessary to feed in the true values of A and E.

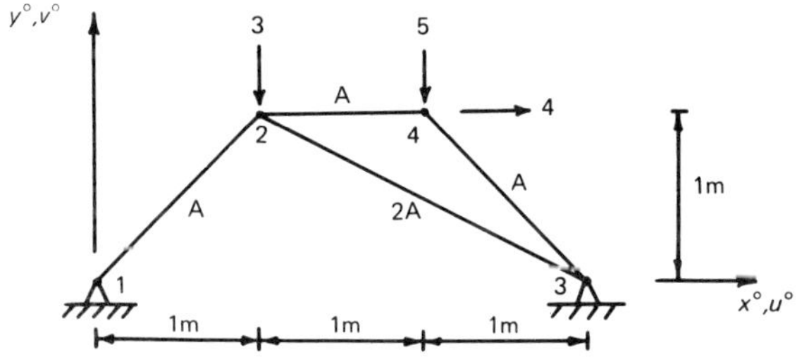

Fig. 1.3.

Number of nodes = 4
Number of members = 4
Number of suppressed displacements = 4

Positions of suppressed displacements

Position 1 = 1 Position 2 = 2 Position 3 = 5 Position 4 = 6

Nodal coordinates

x_i°	y_i°
0	0
1	1
3	0
2	1

Details of members

i	j	A	E
1	2	1	1
2	4	1	1
2	3	2	1
3	4	1	1

Vector of nodal forces

0 0 0 –3 0 0 4 –5

OUTPUT

The eight nodal displacements in global directions $u_1^\circ, v_1^\circ, u_2^\circ, v_2^\circ, \ldots u_4^\circ, v_4^\circ$

Forces in each member

Member	Forces
1	$F_{1-2} = -3.3$
2	$F_{2-4} = -1.0$
3	$F_{2-3} = -1.49$
4	$F_{3-4} = -7.07$

where the subscripts $i - j$ represent the nodes defining the member, and the negative sign denotes compression.

Example 1.2

Determine the member forces in the plane pin-jointed truss shown in Fig. 1.4. For all members

$$A = 1\text{E-}3 \text{ m}^2$$

and

$$E = 2\text{E}8 \text{ kN/m}^2.$$

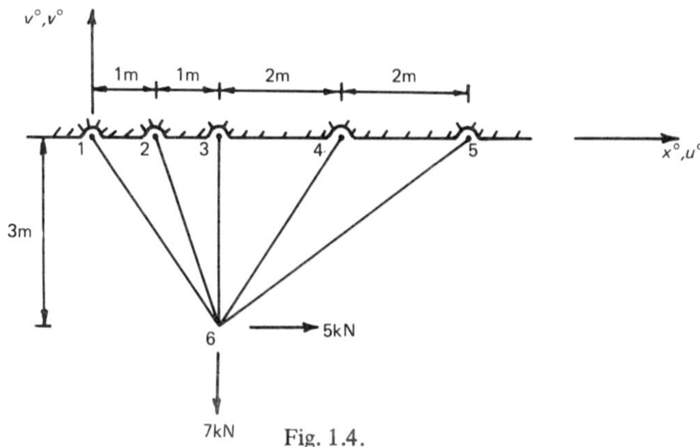

Fig. 1.4.

The data was as follows:

	6		5		10						
	1	2	3	4	5	6	7	8	9	10	
		x_i°			y_i°						
		0			0						
		1			0						
		2			0						
		4			0						
		6			0						
		2			−3						
	i	j	A	E							
	1	6	1E-3	2E8							
	2	6	1E-3	2E8							
	3	6	1E-3	2E8							
	6	4	1E-3	2E8							
	6	5	1E-3	2E8							
0	0	0	0	0	0	0	0	0	0	5	−7

Answers

Node	u°(m)	v°(m)
1	0	0
2	0	0
3	0	0
4	0	0
5	0	0
6	7.58E-5	−3.27E-5

Member	Force (kN)
1-6	3.84
2-6	3.48
3-6	2.18
6-4	−0.82
6-5	−1.64

2

Bending moments in beams

This program, (Appendix 2), can determine the nodal displacements and bending moments of statically determinate and statically indeterminate beams, including continuous beams. The beams can have different sectional properties, and the loading can be a combination of concentrated, distributed and hydrostatic.

Each element has a node at each end, with two degrees of freedom per node, as shown in Fig. 2.1.

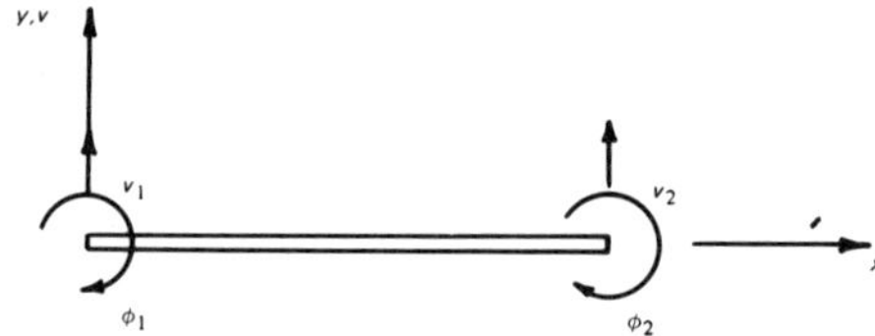

Fig. 2.1.

The displacement positions are calculated as follows:

$$v_i \equiv 2 \times i - 1$$

$$\phi_i \equiv 2 \times i$$

2.1 DATA

The data should be fed in as follows:

(1) Number of elements $= LS$
(2) Number of suppressed displacements $= NF$
(3) Number of concentrated loads $= NC$
(4) *Positions of suppressed displacements*
  FOR $i = 1$ To NF
  → INPUT NS_i
(5) Elastic modulus $= E$

(6) If there is any hydrostatic load feed 1, otherwise feed 0. (that is, $NL = 1$ or $NL = 0$, respectively).

(7) If $NL = 1$ ignore (7)

Member details (No hydrostatic load)

FOR $I = 1$ To LS
(Feed in the members/elements from left to right)
Input SA — Input 2nd moment of area for member
Input XL — Input elemental length
Input UD — Input value of distributed load/length ($+^{ve}$ if in y direction)

(8) If $NL = 0$ ignore (8)

Member details (hydrostatic load)

FOR $I = 1$ To LS
(Feed in the members/elements from left to right)
Input SA — Input 2nd moment of area for element
Input XL — Input elemental length
Input WA — Value of distributed load on left), see Fig. 2.2
Input WB — Value of distributed load on the right) see Fig. 2.2
Input AS — The distance of WA from the left node)

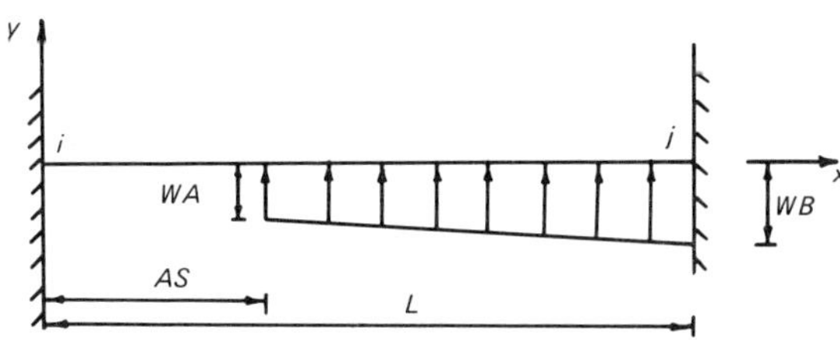

Fig. 2.2

(9) *Position and value of concentrated loads and couples*

FOR $i = 1$ To NC
Input 'Displacement' position of load or couple.
Input value of load or couple.

2.2 OUTPUT

The nodal displacements $v_1, \phi_1, v_2, \ldots v_n, \phi_n,$

where $n = LS + 1$

The nodal bending moments

Element	Nodes defining each element		Nodal bending moments
1	i	j	BM_i
	i	j	BM_j
2	j	k	BM_j
.	j	k	BM_k
.	k		BM_k
.	k		
.			
.			
LS			

Example 2.1

Determine the nodal bending moments for the beam shown in Fig. 2.3. The value of the second moment of area for element 2-3 is twice the value for element 1-2.

As only the moments are required, it will be convenient to assume $I = E = 1$. If, however, the displacements are required, it will be necessary to feed in the true values for E and I.

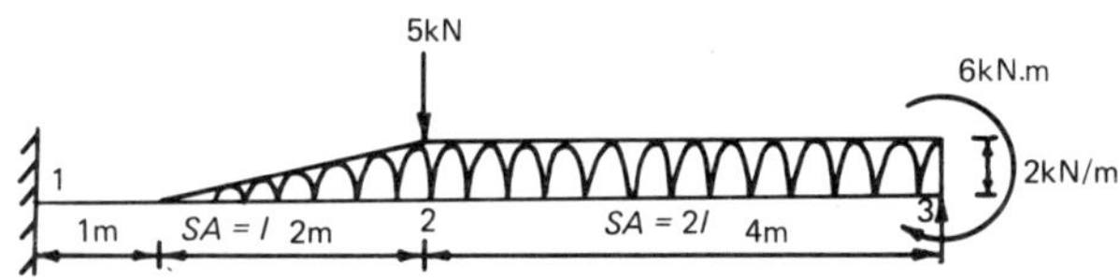

Fig. 2.3.

The data is as follows:

$LS = 2$
$NF = 3$
$NC = 2$

Positions of suppressed displacements

Position 1 = 1
Position 2 = 2
Position 3 = 5

$E = 1$
$NL = 1$

Member details (Hydrostatic load data for this example)

SA	*XL*	*WA*	*WB*	*AS*
1	3	0	–2	1
2	4	–2	–2	0

Position and value of concentrated loads and couples

Position 1 = 3
Value of load = –5
Position 2 = 6
Value of couple = 6 (clockwise $+^{ve}$)

OUTPUT
The displacements v_1, ϕ_1, v_2, ϕ_2, v_3 and ϕ_3

Nodal moments

$M_{1-2} = 11.31$ kN.m – Node 1
$M_{2-1} = -9.06$ kN.m) Member 1
) – Node 2
$M_{2-3} = -9.06$ kN.m) Member 2
$M_{3-2} = \ 6.0$ kN.m – Node 3

where the subscripts $i - j$ are the nodes defining the member. (Hogging moments are $+^{ve}$.)

Example 2.2
Determine the nodal moments for the continuous beam shown in Fig. 2.4.

$E = 2E8$ kN/m^2

$I = 1E\text{-}7$ m^4

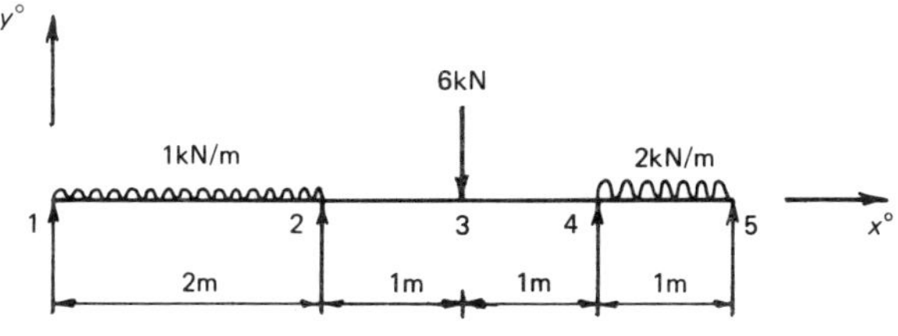

Fig. 2.4.

The data was as follows

4		4		1
1	3	7	9	
2E8				
0				

I	*L*	*UDL*
1E-7	2	–1
1E-7	1	0
1E-7	1	0
1E-7	1	–2
5	–6	

Answers

Node	v(m)	ϕ(rads)
1	0	–1.136E-3
2	0	0.0189
3	–0.0213	–6.629E-4
4	0	–0.0163
5	0	6.061E-3

Member	Node	Moment (kN.m)
1-2	1	0
1-2	2	1.068
2-3	2	1.068
2-3	3	–1.852
3-4	3	–1.852
3-4	4	1.227
4-5	4	1.227
4-5	5	0

3

Bending moments in rigid-jointed plane frames

The program (Appendix 3), can calculate the nodal displacements and bending moments in rigid-jointed plane frames. The elements of the frame can have different sectional properties, and the loading can be a combination of concentrated, distributed and hydrostatic.

Each element has end nodes, and there are three degrees of freedom per node, as shown in Fig. 3.1.

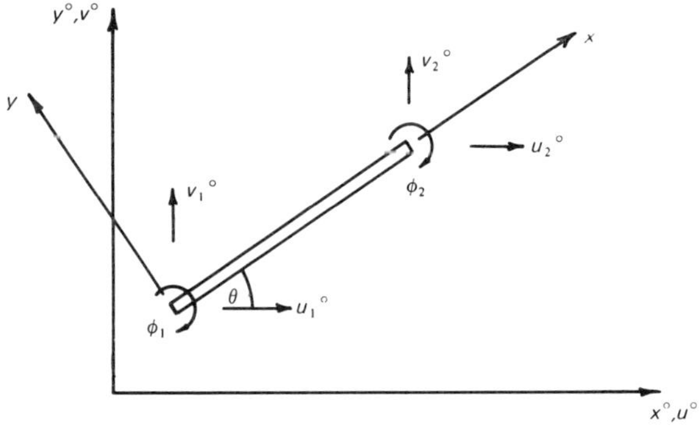

Fig. 3.1.

Displacement positions are calculated as follows:

Deflection position corresponding to $u_i^{\circ} \equiv 3 \times i - 2$
Deflection position corresponding to $v_i^{\circ} \equiv 3 \times i - 1$
Deflection position corresponding to $\phi_i \equiv 3 \times i$

3.1 DATA

The data should be fed in as follows:

(1) Number of nodal points = NJ

(2) Number of elements = MS

(3) Number of suppressed displacements = NF

(4) Number of concentrated loads = NC

(5) Nodal coordinates
FOR $i = 1$ To NJ
→Input x_i°, y_i°

(6) Positions of suppressed displacements
FOR $i = 1$ To NF
→Input NS_i

(7) Elastic modulus = E

(8) If there is any hydrostatic load, feed 1, otherwise feed 0, (that is, $HY = 1$ or $HY = 0$, respectively).

(9) If $HY = 1$, ignore (9).
FOR $I = 1$ To MS
Input i – Input i node of element
Input j – Input j node of element
Input SA – Input 2nd moment of area of element
Input A – Input cross-sectional area of element
Input UD – Input value of uniformly distributed load/length
→($+^{ve}$ if in y direction) – Local axes

(10) If $HY = 0$, ignore (10)
FOR $I = 1$ To MS
Input i – Input i node of element
Input j – Input j node of element
Input SA – Input 2nd moment of area of element
Input A – Input cross-sectional area of element
Input WA – Input value of distributed on 'left') see Fig. 2.2.
Input WB – Input value of distributed load on 'right') Local axes
→Input AS – Distance of WA from the left node

(11) Position and value of concentrated loads and couples
FOR $i = 1$ To NC
→Input Position of load and value of load. (If the load is a couple, then a clockwise direction is taken as positive) – global axes.

3.2 OUTPUT

The nodal displacements $u_1^\circ, v_1^\circ, \phi_1, u_2^\circ, v_2^\circ, \ldots u^\circ_{NJ}, v^\circ_{NJ}, \phi_{NJ}$

Nodal moments and axial forces

Element	Nodes defining element	Axial Force	Moment
1	$i-j$	F_1	M_{i-j}
	$i-j$	F_1	M_{j-i}
2	$k-$	F_2	M_{j-k}
.	$k-$		
.			
.			
MS			

Example 3.1

Determine the nodal bending moments for the rigid-jointed skew frame shown in Fig. 3.2

As only the moments are required, it will be convenient to assume $A/1000 = I = E = 1$. If, however, the nodal displacements are required, it will be necessary to find in the true values of A, I and E, where I is the second moment of area for the inclined members.

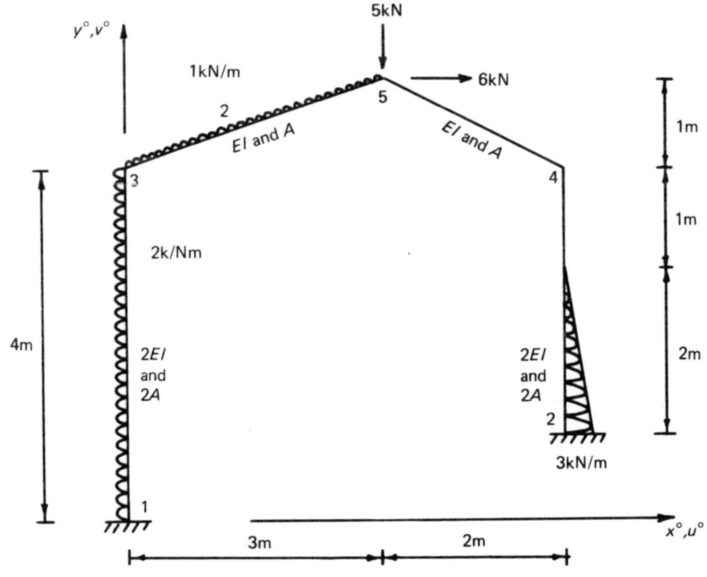

Fig. 3.2.

The data is as follows:

$NJ = 5$
$MS = 4$
$NF = 6$
$NC = 2$

Nodal coordinates

x_i°	y_i°
0	0
5	1
0	4
5	4
3	5

Positions of suppressed displacements

Position 1 = 1
Position 2 = 2
Position 3 = 3
Position 4 = 4
Position 5 = 5
Position 6 = 6

$E = 1$
$HY = 1$

Member details

i	j	SA	A	WA	WB	AS
1	3	2	2000	−2	−2	0
3	5	1	1000	−1	−1	0
5	4	1	1000	0	0	0
4	2	2	2000	0	−3	1

Positions and values of concentrated loads

Position 1 = 13
Value = 6
Position 2 = 14
Value = −5

OUTPUT

The displacements $u_1^\circ, v_1^\circ, \phi_1, u_2^\circ, v_2^\circ, \phi_2, u_3^\circ, v_3^\circ, \phi_3, u_4^\circ, v_4^\circ, \phi_4, u_5^\circ, v_5^\circ$ and ϕ_5

Moments and Forces

Member	Nodes defining member	Axial Force	Nodal moments (kN.m)
1	1-3	–0.87	$M_{1-3} = 11.6$
	1-3	–0.87	$M_{3-1} = -2.57$
2	3-5	–0.71	$M_{3-5} = -2.57$
	3-5	–0.71	$M_{5-3} = 0.28$
3	5-4	–9.86	$M_{5-4} = 0.28$
	5-4	–9.86	$M_{4-5} = 7.08$
4	4-2	–7.13	$M_{4-2} = 7.08$
	4-2	–7.13	$M_{2-4} = -13.3$

Example 3.2

Determine the nodal moments for the rigid-jointed plane frame shown in Fig. 3.3, which is firmly fixed at nodes 1 and 4 and firmly pinned at node 7.

$E = 2E8$ kN/m^2
$I = 1E\text{-}7$ m^4
$A = 1E\text{-}3$ m^2

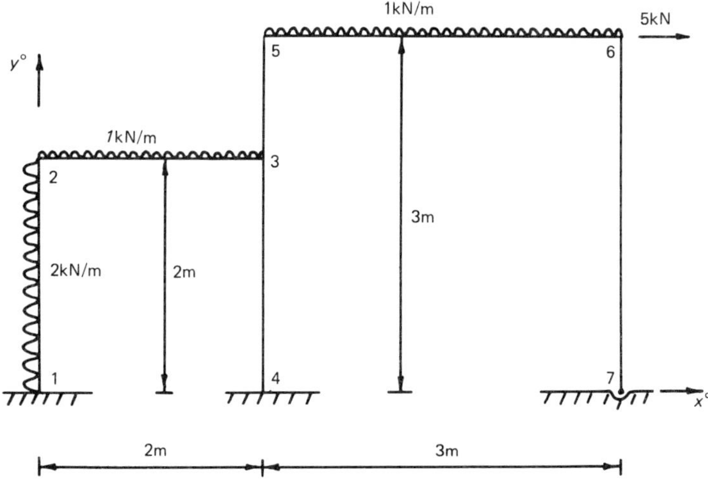

Fig. 3.3.

The data was as follows:

7	6	8	1

x_i°	y_i°
0	0
0	2
2	2
2	0
2	3
5	3
5	0

1	2	3	10	11	12	19	20
	2E8		0				

i	j	I	A	UDL
1	2	1E-7	1E-3	–2
2	3	↓	↓	–1
3	4			0
5	3			0
5	6			–1
6	7			0
16	5			

Answers

Member	Node	Axial Force (kN)	Moment (kN m)
1-2	1	2.532	4.86
1-2	2	2.532	–2.773
2-3	2	1.817	–2.773
2-3	3	1.817	4.290
3-4	3	–4.788	1.741
3-4	4	–4.788	–3.344
5-3	5	–0.256	1.810
5-3	3	–0.256	–2.549
5-6	5	4.359	–1.810
5-6	6	4.359	1.923
6-7	6	–2.744	1.923
6-7	7	–2.744	0

4

Forces in pin-jointed space trusses

This program, (Appendix 4), is meant for determining nodal displacements and member forces in a three dimensional pin-jointed space truss subjected to concentrated loads at its nodes or pin-joints.

Each member of the truss possesses only axial stiffness and has end nodes with three degrees of freedom per node, as shown in Fig. 4.1.

The members of the truss can have different values of cross-sectional area and elastic modulus.

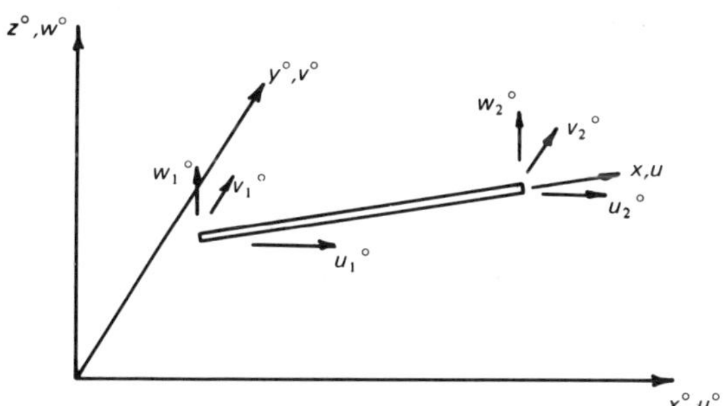

Fig. 4.1 – Three dimensional rod.

The displacement positions can be obtained as follows:

Displacement position $u_i^\circ \equiv 3 \times i - 2$
Displacement position $v_i^\circ \equiv 3 \times i - 1$
Displacement position $w_i^\circ \equiv 3 \times i$

4.1 DATA

The data should be fed in as follows:

(1) Number of pin-joints or nodes $= NJ$
(2) Number of members $= MS$
(3) Number of suppressed displacements $= NF$

(4) *Nodal coordinates*

FOR $i = 1$ To NJ

→Input x_i°, y_i°

(5) Positions of suppressed displacements

FOR $i = 1$ To NF

→Input NS_i

(6) *Member details*

FOR $I = 1$ To MS

Input i — Input i node for member

Input j — Input j node for member

Input A — Input cross-sectional area for member

→Input E — Input elastic modulus for member

(7) *Vector of externally applied loads*

FOR $i = 1$ To $N3$

→Input $(\text{load})_i$

where $N3 = 3 \times NJ$

4.2 OUTPUT

The nodal displacements $u_1^{\circ}, v_1^{\circ}, w_1^{\circ}, u_2^{\circ}, v_2^{\circ}, w_2^{\circ}, \ldots u_{NJ}^{\circ}, v_{NJ}^{\circ}, w_{NJ}^{\circ}$.

Axial forces in members

Member	Nodes defining member	*Force*
1	$i - j$	F_{i-j}
2		
.		
.		
.		
MS		

N.B. Tensile forces are positive.

Example 4.1

Determine the forces in the three dimensional pin-jointed truss shown in Fig. 4.2. All members are of constant AE.

$NJ = 5$

$MS = 4$

$NF = 12$

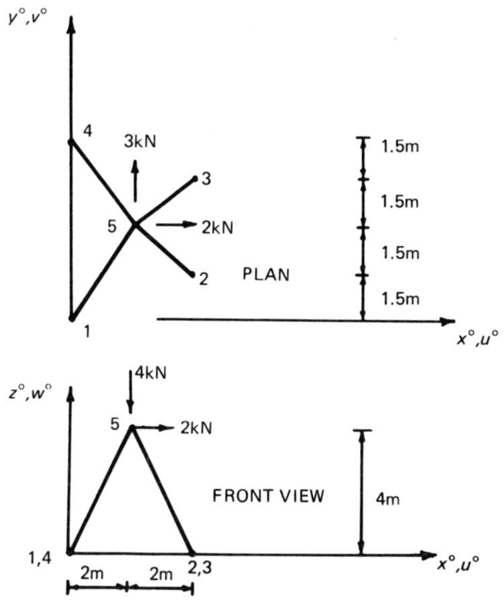

Fig. 4.2 – Three dimensional truss.

Nodal coordinates

x_i°	y_i°	z_i°
0	0	0
4	1.5	0
4	4.5	0
0	6	0
2	3	4

Suppressed displacement positions

Position 1 = 1
Position 2 = 2
Position 3 = 3
Position 4 = 4
Position 5 = 5
Position 6 = 6
Position 7 = 7
Position 8 = 8
Position 9 = 9
Position 10 = 10
Position 11 = 11
Position 12 = 12

Member details

i	j	A	E
1	5	1	1
2	5	1	1
3	5	1	1
4	5	1	1

Vector of externally applied loads

0	0	0
0	0	0
0	0	0
0	0	0
2	3	–4

OUTPUT

The nodal displacements u_1°, v_1°, w_1°, u_2°, v_2°, w_2°, u_3°, v_3°, w_3°, u_4°, v_4°, w_4°, u_5°, v_5° and w_5°

Forces in members

Member	Nodes defining member	*Force*
1	1-5	1.96
2	2-5	–1.08
3	3-5	–3.64
4	4-5	–1.96

The $-^{ve}$ sign denotes compression.

Example 4.2

Determine the member forces in the pin-jointed space truss of Fig. 4.3, which is firmly pinned at nodes 1 to 4. It may be assumed that $AE = 1$.

The data was as follows:

6	7	12
x_i°	y_i°	z_i°
0	–0.5	0
0	0.5	0
0	0.5	–1
0	–0.5	–1
1	0	–0.5
1.5	0	–2

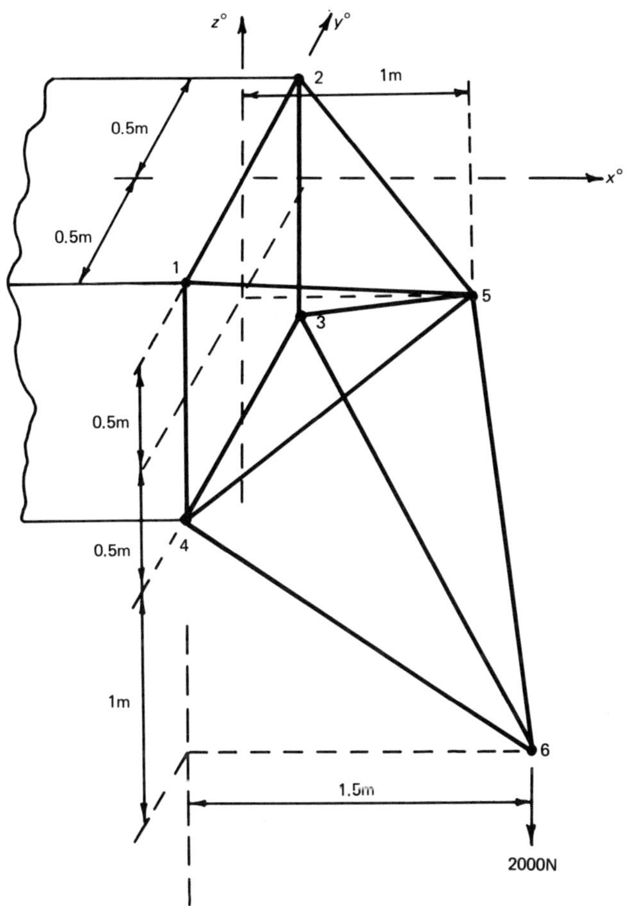

Fig. 4.3.

1	2	3	4	5	6	7	8	9	10	11	12
i		*j*			*A*			*E*			
1		5			1			1			
2		5									
3		5									
5		4									
5		6									
6		3									
4		6									
0	0	0	0	0	0	0	0	0	0	0	0
0	0	0	0	0	–2000						

Answers

Node	u°	(m) v°	w°
1	0	0	0
2	0	0	0
3	0	0	0
4	0	0	0
5	393.7	0	–4724
6	–9637	0	–12585

Member	Force (N)
1-5	1837
2-5	1837
3-5	–1312
5-4	–1312
5-6	2711
6-3	– 535
4-6	– 535

5

Free vibration of plane pin-jointed trusses

The program, (Appendix 5), can determine the natural frequencies and eigenmodes of plane pin-jointed trusses of the type discussed in Chapter 1.

The equation of motion for free vibration, without damping, can be reduced to the determinant of (5.1).

$$|[K] - \omega^2 [M]| = 0 \tag{5.1}$$

where,

$[K]$ = the stiffness matrix corresponding to free displacements only.
$[M]$ = the mass matrix corresponding to free displacements only.
ω = circular frequency.

As the problems in the present text are confined to constrained and over-constrained structures, it will be convenient to invert $[K]$ and pre-multiply it into (5.1) to give

$$|\lambda[I] - [K]^{-1} [M]| = 0 \tag{5.2}$$

where $\lambda = 1/\omega^2$ = eigenvalue
and $n = \omega/2\pi$ = frequency

Solution of (5.2) can be carried out by a number of methods, including the power method.

In all the vibration programs, the precision factor D has been set to 0.001. This is equivalent to 0.1%, and if a greater or smaller precision is sought, the value of the constant D in the program can be altered.

It should be noted that no attempt was made to seek any precision of the eigenvectors, and because of this, the precision of the eigenvectors is likely to be less than that of the eigenvalues.

It should also be noted that the computer program cannot determine eigenvalues which are of equal value, that is, the magnitudes of the eigenvalues should be distinct.

5.1 DATA

The data should be fed in as follows:

(1) Number of nodal points or pin-joints $= NN$

(2) Number of members $= LS$

(3) Number of suppressed displacements $= NF$

(4) Number of frequencies $= M1$

where $M1 \leqslant N$
and $N = 2 \times NN - NF$

(5) Positions of suppressed displacements
FOR $i = 1$ To NF
→Input NS_i in ascending order

(6) *Nodal coordinates*
FOR $i = 1$ To NN
→Input $x_i°, y_i°$

(7) *Member details*
FOR $I = 1$ To LS
Input i — Input i node for member
Input j — Input j node for member
Input A — Input cross-sectional area for member
Input E — Input elastic modulus for member
→Input ρ — Input density for member

(8) Number of concentrated masses $= NC$

(9) Concentrated masses and their nodal positions
FOR $i = 1$ To NC
→Input nodal position of mass, and value of mass

Example 5.1
Determine the two lowest natural frequencies of vibration for the truss of Example 1.1, given that:

$$\rho = 7860 \text{ kg/m}^3$$
$$E = 2 \times 10^{11} \text{ N/m}^2$$
$$A = 0.0003 \text{ m}^2$$

Concentrated mass at node 2 = 4 kg
Concentrated mass at node 4 = 5 kg

The data is as follows:

$NN = 4$
$LS = 4$
$NF = 4$
$M1 = 2$

Positions of suppressed displacements
As in Chapter 1.

Nodal coordinates
As in Chapter 1.

Member details

i	j	A	E	ρ
1	2	3E-4	2E11	7860
2	4	3E-4	2E11	7860
2	3	6E-4	2E11	7860
3	4	3E-4	2E11	7860

$NC = 2$

Nodal positions and values of concentrated masses
Position 1 = 2
Mass value = 4
Position 2 = 4
Mass value = 5

Results
1st frequency $= n_1 = 187.3$ Hz

$$\begin{Bmatrix} u_2^\circ \\ v_2^\circ \\ u_4^\circ \\ v_4^\circ \end{Bmatrix} = \begin{Bmatrix} 0.299 \\ 0.036 \\ 0.549 \\ 1.0 \end{Bmatrix}$$ 1st eigenmode – See Fig. 5.1.

2nd frequency $= n_2 = 293.7$ Hz

$$\begin{Bmatrix} u_2^\circ \\ v_2^\circ \\ u_4^\circ \\ v_4^\circ \end{Bmatrix} = \begin{Bmatrix} -0.03 \\ 1.0 \\ -0.052 \\ -0.062 \end{Bmatrix}$$ 2nd eigenmode – See Fig. 5.2.

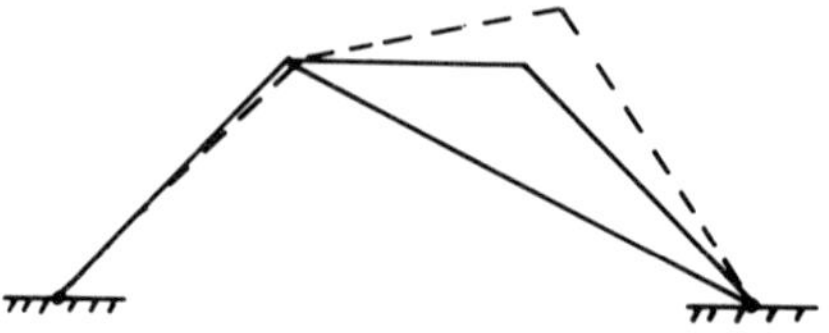

Fig. 5.1

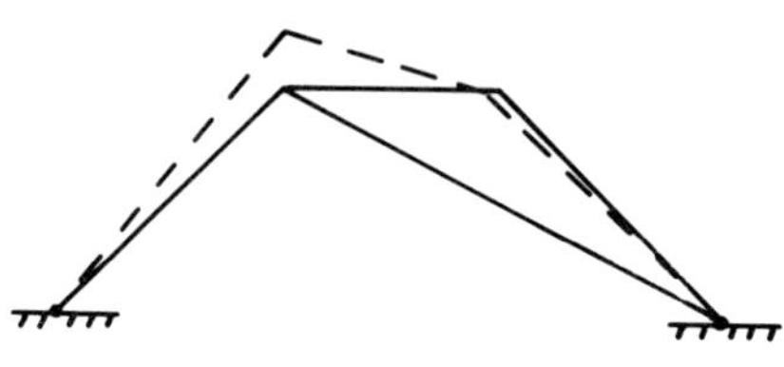

Fig. 5.2

Example 5.2

Determine the two lowest frequencies of vibration for the plane pin-jointed truss of Example 1.2, given that $\rho = 7860$ kg/m^3.

For this case, E must be in units of N/m^2.

Answers

$n_1 = 186.4$ Hz
$u_6^\circ = 1 \qquad v_6^\circ = 0.0933$

$n_2 = 336.2$ Hz
$u_6^\circ = 1.52\text{E}{-3} \qquad v_6^\circ = 1$

6

Free vibration of continuous beams

This program, (Appendix 6), can determine the natural frequencies and eigenmodes of beams composed of elements of different sectional properties. The element is the same as that adopted in Chapter 2, and the equations of motion are the same as that described in Chapter 5.

6.1 DATA

The data should be fed in as follows:

(1) Number of elements = LS
(2) Number of suppressed displacements = NF
(3) Number of frequencies required = $M1$

(4) *Positions of suppressed displacements*
┌FOR i = 1 To NF
└→Input NS_i in ascending order
(5) Elastic modulus = E
(6) Density = ρ

(7) *Element details*
┌FOR i = 1 To LS
│Input i — Input i node for element (left node)
│Input j — Input j node for element (right node)
│Input SA — Input 2nd moment of area for element
│Input CA — Input cross-sectional area for element
└→Input L — Input elemental length
(8) Number of concentrated masses = NC
(9) Position, and value of mass and mass moment of inertia.
┌FOR i = 1 To NC
│Input position
│Input mass
└→Input mass moment of inertia

Example 6.1

Determine the two lowest natural frequencies of vibration of the continuous beam of Chapter 2, where

$$
\begin{aligned}
I &= 2\text{E-}8 \text{ m}^4 \\
A &= 5\text{E-}4 \text{ m}^2 \\
\rho &= 7860 \text{ kg/m}^3 \\
E &= 2\text{E}11 \text{ N/m}^2
\end{aligned}
$$

At node 2, there is a mass of value 11.29 kg with a mass moment of inertia of 0.022 kg.m^2. The cross-sectional area for element 2–3 may be assumed to be $2A$.

The data is as follows:

$LS = 2$
$NF = 3$
$M1 = 2$

Positions of suppressed displacements – As in Chapter 2.

$E = 2\text{E}11$
$\rho = 7860$

Element details

i	j	SA	CA	L
1	2	2E–8	5E–4	3
2	3	4E–8	10E–4	4

$NC = 1$

Position and value of mass and mass moment of inertia

Nodal Position = 2
Value of mass = 11.29
Value of mass moment of inertia = 0.022

Results

$n_1 = 1.234$ Hz

$$\begin{Bmatrix} v_2 \\ \phi_2 \\ \phi_3 \end{Bmatrix} = \begin{Bmatrix} 1 \\ -0.16 \\ 0.507 \end{Bmatrix}$$ See Fig. 6.1.

$n_2 = 5.106$ Hz

$$\begin{Bmatrix} v_2 \\ \phi_2 \\ \phi_3 \end{Bmatrix} = \begin{Bmatrix} -0.425 \\ -0.786 \\ 1 \end{Bmatrix}$$

Fig. 6.1 – 1st Eigenmode.

Example 6.2
Determine the two lowest frequencies of vibration for the continuous beam of Example 2.2, given that

$$\rho = 7860 \text{ kg/m}^3 \text{ and } A = 1\text{E-3 m}^2.$$

E must be in units of N/m².

Answers

$$n_1 = 23.84 \text{ Hz}$$

Eigen mode 1

Node	$v°$	ϕ
1	0	1
2	0	–0.883
3	0.445	0.132
4	0	0.383
5	0	–0.202

$$n_2 = 39.0 \text{ Hz}$$

Eigen mode 2

Node	$v°$	ϕ
1	0	1
2	0	–0.086
3	–0.345	0.144
4	0	–0.418
5	0	0.240

7

Free vibration of rigid-jointed plane frames

This program, (Appendix 7), can determine the natural frequencies and eigenmodes for a rigid-jointed plane frame. The elements of the frame can have different sectional properties, as in Chapter 3. The vibration equation is the same as that described in Chapter 5.

7.1 DATA

The data should be fed in as follows:

(1) Number of nodal points $= NN$
(2) Number of elements $= MS$
(3) Number of suppressed displacements $= NF$
(4) Number of frequencies $= M1$

(5) *Nodal coordinates*
(As in Chapter 3)
(6) Suppressed displacement positions in ascending order
(7) Elastic modulus $= E$
(8) Density $= \rho$

(9) *Member details*

┌FOR $i = 1$ To MS
│Input i – Input i node for element
│Input j – Input j node for element
│Input SA – Input 2nd moment of area for element
└→Input CA – Input cross-sectional area for element

(10) Number of concentrated masses $= NC$

(11) *Positions and values of mass and mass moment of inertia*

┌FOR $i = 1$ To NC
│Input position
│Input mass
└→Input mass moment of inertia

Example 7.1

Determine the two lowest natural frequencies of vibration for the rigid-jointed plane frame shown in Chapter 3, given the following:

$$I = 1.03 \times 10^{-7}\,\mathrm{m}^4$$
$$A = 1.14 \times 10^{-3}\,\mathrm{m}^2$$
$$E = 2 \times 10^{11}\,\mathrm{N/m}^2$$
$$\rho = 7860\,\mathrm{kg/m}^3$$

It may be assumed that node 5 has attached to it a mass of value 33 kg with a mass moment of inertia of 0.132 kg.m^2.

The data is as follows:

$NN = 5$
$MS = 4$
$NF = 6$
$M1 = 2$

Nodal coordinates – As in Chapter 3.

Suspressed displacement positions – As in Chapter 3.

$E = 2E11$
$\rho = 7860$

Member details

i	*j*	*SA*	*CA*
1	3	2.06E-7	2.28E-3
3	5	1.03E-7	1.14E-3
5	4	1.03E-7	1.14E-3
4	2	2.06E-7	2.28E-3

$NC = 1$

Position and value of mass and mass moment of inertia

Nodal position = 5
Value of mass = 33 kg
Value of mass moment of inertia = 0.132 kg.m^2

Results

1st frequency = n_1 = 1.55 Hz

$$\begin{Bmatrix} u_3^\circ \\ v_3^\circ \\ \phi_3 \\ u_4^\circ \\ v_4^\circ \\ \phi_4 \\ u_5^\circ \\ v_5^\circ \\ \phi_5 \end{Bmatrix} = \begin{Bmatrix} 1 \\ 3.08\mathrm{E}{-5} \\ 0.19 \\ 0.82 \\ -1.6\mathrm{E}{-5} \\ 0.325 \\ 0.93 \\ 0.21 \\ -0.091 \end{Bmatrix} = \text{1st eigenmode (see Fig. 7.1).}$$

2nd frequency = 3.67 Hz

$$\begin{Bmatrix} u_3^\circ \\ v_3^\circ \\ \phi_3 \\ u_4^\circ \\ v_4^\circ \\ \phi_4 \\ u_5^\circ \\ v_5^\circ \\ \phi_5 \end{Bmatrix} = \begin{Bmatrix} 0.216 \\ 1.05E\text{-}4 \\ -0.28 \\ -0.617 \\ 1.33E\text{-}4 \\ 0.091 \\ -0.11 \\ 1 \\ 0.302 \end{Bmatrix} = \text{2nd eigenmode (see Fig. 7.2).}$$

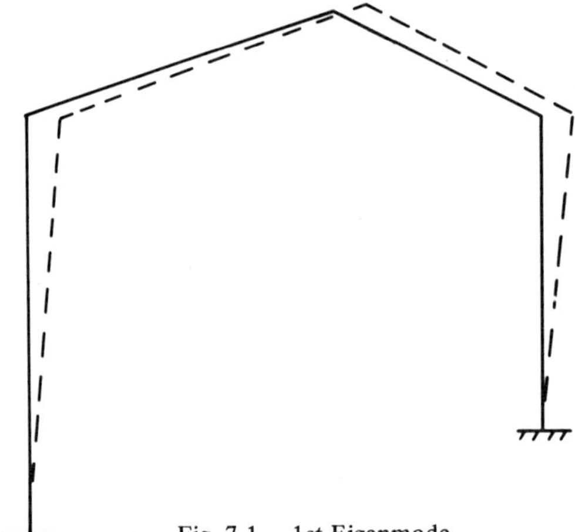

Fig. 7.1 – 1st Eigenmode.

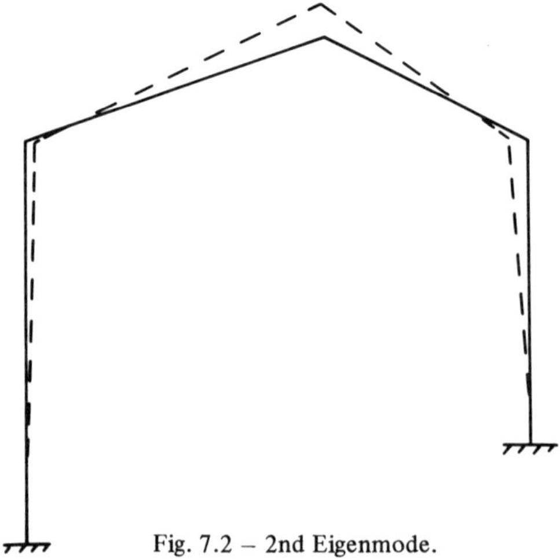

Fig. 7.2 – 2nd Eigenmode.

Example 7.2

Determine the three lowest frequencies of vibration for the rigid-jointed plane frame of Example 3.2, given that $\rho = 7860$ kg/m^3. E must be in units of N/m^2.

Answers

$n_1 = 3.416$ Hz

Eigenmode 1

Node	u°	v°	ϕ
1	0	0	0
2	0.639	1.37E-4	0.16
3	0.639	–1.02E-4	0.296
4	0	0	0
5	1	–8.51E-5	0.296
6	1	–6.23E-5	–7.92E-3
7	0	0	0.562

$n_2 = 9.45$ Hz

Eigenmode 2

Node	u°	v°	ϕ
1	0	0	0
2	–0.616	1.7E-5	–0.267
3	–0.616	–1.6E-4	0.246
4	0	0	0
5	0.103	–2.6E-4	0.948
6	0.104	–2.96E-4	–0.93
7	0	0	1

$n_3 = 15.18$ Hz

Eigenmode 3

Node	u°	v°	ϕ
1	0	0	0
2	0.23	–1.06E-4	0.167
3	0.23	–1.67E-4	–0.336
4	0	0	0
5	–0.302	–2.42E-4	–0.359
6	–0.302	–2.76E-5	–0.676
7	0	0	1

8

Free vibration of pin-jointed space trusses

This program, (Appendix 8), can determine the natural frequencies of vibration and eigenmodes of three dimensional pin-jointed trusses. The members of the truss can have different sectional and material properties, as described in Chapter 4. The vibration solution is similar to that described in Chapter 5.

8.1 DATA

The data should be fed in as follows:

(1) Number of nodes or pin-joints $= NJ$
(2) Number of suppressed displacements $= NF$
(3) Number of members $= MS$
(4) Number of frequencies $= M1$

(5) *Nodal coordiantes*
┌FOR $i = 1$ To NJ
└→Input $x_i^\circ, y_i^\circ, z_i^\circ$

(6) *Member details*
┌FOR $I = 1$ To MS
│Input i – Input i node for member
│Input j – Input j node for member
│Input A – Input cross-sectional area for member
│Input E – Input elastic modulus for member
└→Input ρ – Input density for member

(7) *Suppressed displacements*
┌FOR $i = 1$ To NF
└→Input NS_i in ascending order

(8) Number of concentrated masses = NC
(9) Nodal position and value of mass
┌FOR $i = 1$ To NC
│Input Position of mass
└→Input Value of mass

Example 8.1

Determine the two lowest natural frequencies of vibration for the pin-jointed truss shown in Chapter 4, given the following:

$$A = 0.001 \text{ m}^2$$
$$E = 2 \times 10^{11} \text{ N/m}^2$$
$$\rho = 7860 \text{ kg/m}^3$$

It has attached to node 5 a mass of 30 kg.

The data is as follows:

$NJ = 5$
$NF = 12$
$MS = 4$
$M1 = 2$

Nodal coordinates – As in Chapter 4.

Member details

i	j	A	E	ρ
1	5	0.001	2E11	7860
2	5	0.001	2E11	7860
3	5	0.001	2E11	7860
4	5	0.001	2E11	7860

Suppressed displacement positions – As in Chapter 4.

$NC = 1$

Position and value of mass

Nodal position = 5
Value of mass = 30 kg

Results

1st frequency $= n_1 = 85.97$ Hz

$$\begin{Bmatrix} u_5^\circ \\ v_5^\circ \\ w_5^\circ \end{Bmatrix} = \begin{Bmatrix} 1 \\ 0.24 \\ 0.13 \end{Bmatrix}$$ = 1st eigenmode – see Fig. 8.1.

2nd frequency $= n_2 = 98.30$ Hz

$$\begin{Bmatrix} u_5^\circ \\ v_5^\circ \\ w_5^\circ \end{Bmatrix} = \begin{Bmatrix} -1.16\text{E-}3 \\ 1 \\ 2.67\text{E-}3 \end{Bmatrix}$$ = 2nd eigenmode – see Fig. 8.2.

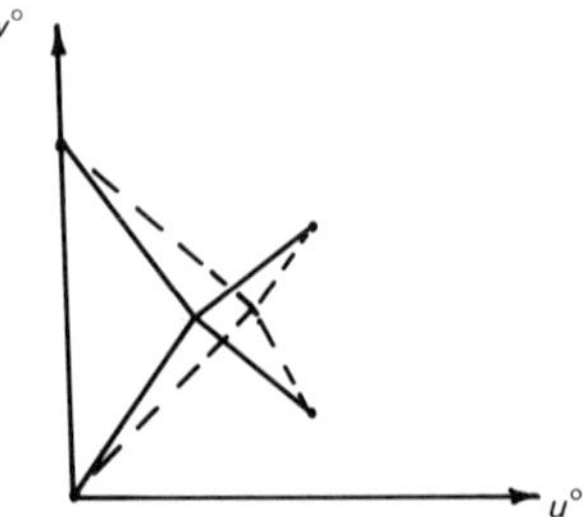

Fig. 8.1 – First eigenmode.

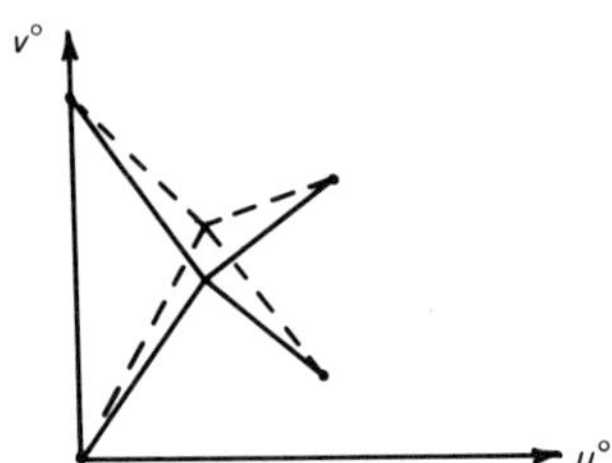

Fig. 8.2 – Second eigenmode.

Example 8.2

Determine the two lowest frequencies of vibration for the pin-jointed space truss of Example 4.2, given that

$$\rho = 7860 \text{ kg/m}^3$$
$$E = 2\text{E}11 \text{ N/m}^2$$
$$A = 1\text{E-3 m}^2$$

Answers

$n_1 = 166.2$ Hz

Node	u°	v°	w°
5	–0.0175	0.025	0.321
6	0.665	1.0	0.783

$n_2 = 172.1$ Hz

Node	u°	v°	w°
5	–0.022	1.48E–4	0.410
6	0.8497	–3.47E–4	1

9

Forces and moments in rigid-jointed space frames

This program (Appendix 9), is intended for analysing 3-dimensional rigid-jointed space frames.

Each node has six local degrees of freedom, (u, v, w, θ_x, θ_y and θ_z), and these are shown with the six global degrees of freedom (u°, v°, w°, $\theta_x{}^\circ$, $\theta_y{}^\circ$, and $\theta_z{}^\circ$), in Fig. 9.1.

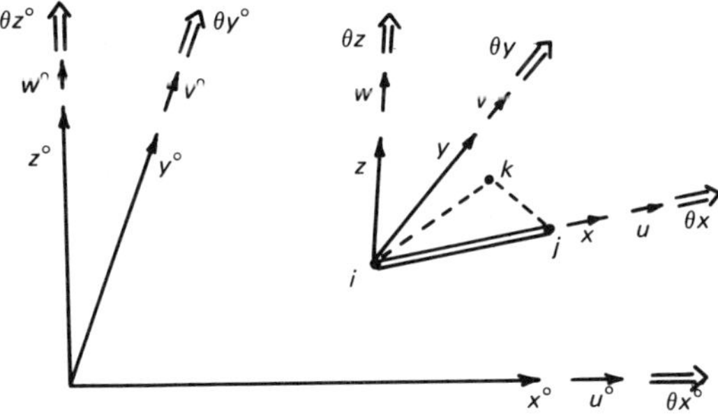

Fig. 9.1 – General case of the one-dimensional member.

The global axes are $x^\circ - y^\circ - z^\circ$, and the local axes are $x - y - z$. The node k is in the local $x - y$ plane, so that it defines one of the principal planes of bending, as shown in Fig. 9.2. The other principal plane is the local $x - z$ plane and is perpendicular to the $x - y$ principal plane.

The directions of the couples and the rotations are shown according to the right-hand screw rule.

The elemental stiffness matrix in local co-ordinates is given in [15] and [16].

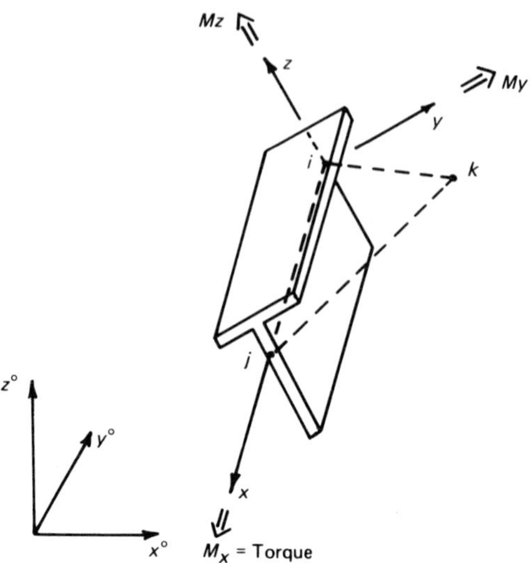

Fig. 9.2 – Principal planes $x - y$ and $x - z$.

9.1 DATA

This should be fed in as follows:

- NJ – number of actual joints or nodes
- NF – number of suppressed displacements
- IM – number of imaginary nodes. These are required in addition to the actual nodes if it is not possible to define the principal $x - y$ planes of the members with the actual nodes. (see later.)
- ML – number of different types of member. Each particular member type has the same sectional and material properties.
- MS – number of members.

Global co-ordinates of nodes

$i = 1(1)NJ + IM$
$\rightarrow x_i^\circ \; y_i^\circ \; z_i^\circ$

Details of member types

$i = 1(1)ML$
$\rightarrow E_i \; G_i \, AR_i \; IX_i \; IY_i \; IZ_i$

where,

E = elastic modulus
G = rigidity modulus
AR = cross-sectional area
IX = torsional constant
IY = second moment of area about the (local) principal plane $x - y$
IZ = second moment of area about the (local) principal plane $x - z$.

Nodes defining each member

┌ $ME = 1(1)MS$
└→ i_{ME} j_{ME} k_{ME}

where,

$i =$ the node which is of the origin of the local co-ordinate system.
$j =$ the node which is at the end of the member, so that $i \rightarrow j$ represents the x direction.
$k =$ a node in the (local) principal plane $x - y$, as shown in Fig. 9.2. In many cases, k can be an existing node in the structure, which may conveniently define the $x - y$ principal plane of the member. If there is no such convenient node, then an imaginary node must be used. Imaginary nodes are used for defining principal planes of bending and do not increase the size of the stiffness matrix, nor the size of the half band width.

If there is more than one type of member (that is, $ML > 1$), then, immediately after feeding in i, j and k, the member type should be fed in, as follows:

If $ML > 1$,

┌ $ME = 1(1)MS$
│ i j k
└→ Member type (a number 1, 2, 3, etc.)

NC — number of nodes with concentrated loads

┌ $i = 1(1)NC$
│ PW_i — nodal position
│ Force in x° direction
│ Force in y° direction
│ Force in z° direction
│ Couple in x° direction)
│ Couple in y° direction) according to the right-hand screw rule
└→ Couple in z° direction)

Positions of suppressed displacements

┌ $i = 1(1)NF$
└→ NS_i — 'Displacement' position of each suppressed displacement.

Example (on method of calculating NS_i)
If in a structure, node 8 is fixed and node 11 pinned, then the number of suppressed displacements corresponding to node 8 is 6 and the number of suppressed displacements corresponding to node 11 is 3, so that $NF = 9$.

The positions of the suppressed displacements, together with their calculation are as follows:

Position 1 $= 6 \times 8 \quad - 5 = 43$
Position 2 $= 6 \times 8 \quad - 4 = 44$
Position 3 $= 6 \times 8 \quad - 3 = 45$
Position 4 $= 6 \times 8 \quad - 2 = 46$
Position 5 $= 6 \times 8 \quad - 1 = 47$
Position 6 $= 6 \times 8 \quad\quad = 48$
Position 7 $= 6 \times 11 - 5 = 61$
Position 8 $= 6 \times 11 - 4 = 62$
Position 9 $= 6 \times 11 - 3 = 63$

The output is as follows:

Global displacements

$$u_1^\circ\ v_1^\circ\ w_1^\circ\ \theta_{x1}^\circ\ \theta_{y1}^\circ\ \theta_{z1}^\circ\ u_2^\circ\ v_2^\circ \ldots$$

$$\theta_{xNJ}^\circ\ \theta_{yNJ}^\circ\ \theta_{zNJ}^\circ$$

Axial forces and moments

┌$ME = 1(1)MS$
└→Axial force $T\ BM_y\ BM_z$ (for member $i - j - k$)

where,

T = torque)
BM_y = bending moment about $x - y$ plane) local axes – see Figs. 9.3 and 9.4
BM_z = bending moment about $x - z$ plane)

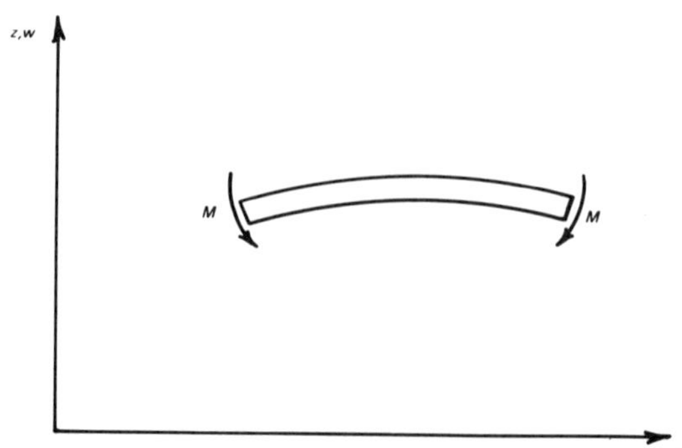

Fig. 9.3 – Positive bending moment about $x - y$ plane.

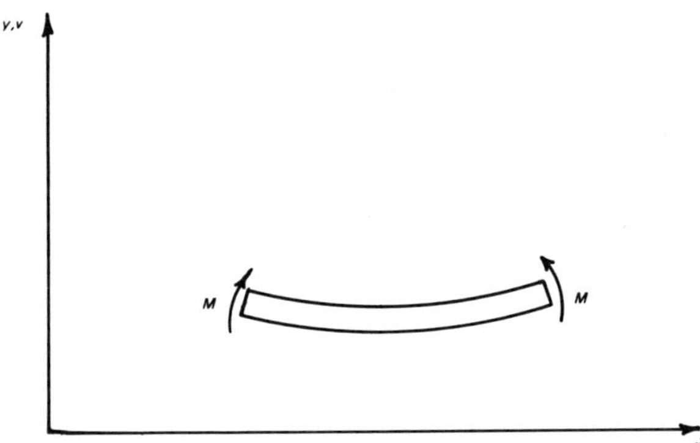

Fig. 9.4 – Positive bending moment about $x - z$ plane.

Example 9.1

Determine the nodal displacements moments and forces in the rigid-jointed space frame shown in Fig. 9.5. The frame is similar to that of [15,p. 35], but the loading is not the same. All members are of circular section with the following properties:

$A = 25.13$ in^2 $I_{xx} = J = 125.7$ in^4 $I_{xy} = I_{xz} = 62.83$ in^4
$E = 13{,}500$ tonf/in^2 $G = 5{,}500$ tonf/in^2.
Length of each member = 120 in

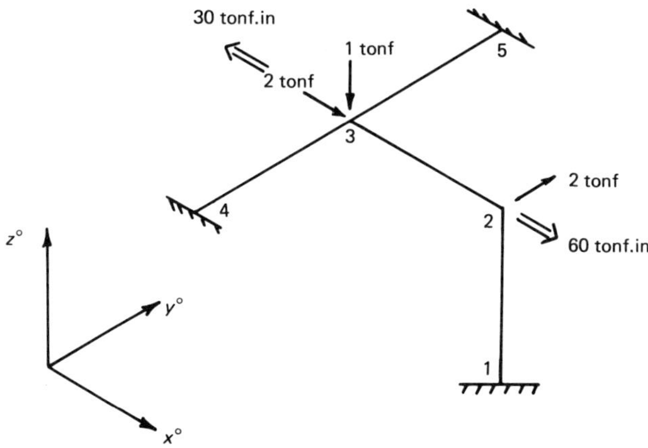

Fig. 9.5 – Rigid-jointed space frame.

The data should be fed in as follows:

NJ = 5

NF = 18 (There are three fixed nodes with 6 degrees of freedom per node).

IM = 0 As the members are of circular cross-section, each cross-section has an infinite number of principal axes of bending, thus, for this particular case, no imaginary nodes are required to define the $x - y$ axes. (Almost any node, apart from *i* and *j* will be suitable to be used as a *k* node.)

ML = 1 There is only one 'type' of member.

MS = 4 There are four members.

Global co-ordinates of nodes

x°	y°	z°	
120	0	0)	
120	0	120)	
0	0	120)	Any suitable point can be taken as the origin,
0	–120	120)	but the system *must be the normal right*
0	120	120)	*handed system for* x°, y° and z°.

E = 13,500)
G = 5,500)
AR = 25.13) Member properties
IX = 125.7)
IY = 62.83)
IZ = 62.83)

Nodes defining each member

i	*j*	*k*	
1	2	3	($x - y$ is a vertical plane)
3	2	1	($x - y$ is a vertical plane)
4	3	2	($x - y$ is a horizontal plane)
5	3	2	($x - y$ is a horizontal plane)

NC = 2 (There are two nodes which have externally applied concentrated loads.)

Nodal positions and 'values' of concentrated loads

Nodal position 1 = 2

0 2 0 (Forces) 60 0 0 (Couples)

Nodal position 2 = 3

2 0 –1 (Forces) –30 0 0 (Couples)

Displacement positions of suppressed displacements

1	2	3	4	5	6	(Corresponding to node 1)
19	20	21	22	23	24	(Corresponding to node 4)
25	26	27	28	29	30	(Corresponding to node 5)

Results

Displacements are zero at nodes 1, 4 and 5. Values for deflections at nodes 2 and 3 are:

$$u_2^\circ = 0.125 \text{ ins}, \; v_2^\circ = 0.364 \text{ ins}, \; w_2^\circ = -1.06\text{E-4 ins}$$

$$u_3^\circ = 0.125 \text{ ins}, v_3^\circ = 1.08\text{E-4 ins}, w_3^\circ = -0.0595 \text{ ins}$$

Moments and forces in units of tonf and in.

Member	Node	Axial Force	Torque	BM_{x-y}	BM_{x-z}
1-2-3	1	–.299	19.46	98.41	–35.76
1-2-3	2	–.299	19.46	–68.35	27.35
3-2-1	3	–.526	– 8.35	–53.78	– 8.51
3-2-1	2	–.526	– 8.35	19.46	27.35
4-3-2	4	+.305	– 4.256	–11.44	57.67
4-3-2	3	+.305	– 4.256	1.861	–71.11
5-3-2	5	–.305	4.256	30.62	30.78
5-3-2	3	–.305	4.256	–40.21	–17.33

Example 9.2

Determine the nodal displacements and moments for the cantilever shown in Fig. 9.6. Both members are of solid rectangular cross-section with the following properties:

$E = 2 \times 10^{11}$ N/m^2 $G = 7.7 \times 10^{10}$ N/m^2.
Depth = 0.1 m and width = 0.08 m for 1-2;
Depth = 0.06 m and width = 0.04 m for member 2-3;
Length 1-2 = length 2-3 = 1m.

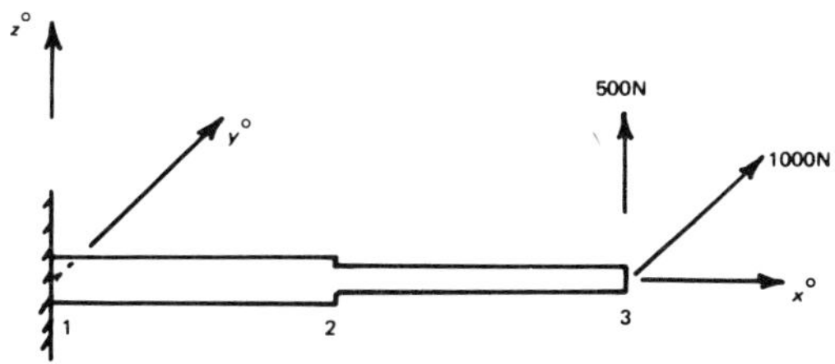

Fig. 9.6 – Cantilever.

The data is as follows:

$NJ = 3$
$NF = 6$
$IM = 1$ It will be necessary to have at least one imaginary node to define the $x - y$ planes. This node can be at almost any position which will help to define either a vertical or horizontal plane, that is, a principal plane.
$ML = 2$ There are two different types of member, as one member has different cross-sectional dimensions to the other.
$MS = 2$

Global co-ordinates

x°	y°	z°	
0	0	0	
1	0	0	
2	0	0	
1	1	0	← This is the imaginary node. Its position will define the $x - y$ principal planes as horizontal, that is, the $x - z$ principal planes will be vertical for this case. (*N.B.* These could have been vice-versa if y° were zero and z° had a value.)

$E = 2 \times 10^{11}$)
$G = 7.7 \times 10^{10}$)
$AR = 8.0 \times 10^{-3}$) member type (1)
$IX = 8.759 \times 10^{-6}$)
$IY = 6.667 \times 10^{-6}$)
$IZ = 4.267 \times 10^{-6}$)

$E = 2 \times 10^{11}$)
$G = 7.7 \times 10^{10}$)
$AR = 2.4 \times 10^{-3}$) member type (2)
$IX = 7.512 \times 10^{-7}$)
$IY = 7.2 \times 10^{-7}$)
$IZ = 3.2 \times 10^{-7}$)

Other member details

i j k and member types
1 2 4

member type = 1
2 3 4

member type = 2

(that is, member 1-2 is member type (1) and member 2-3 is member type (2)).

$NC = 1$

Values and displacement positions of nodal forces

Nodal position = 3

$$\underbrace{0 \quad 1000 \quad 5000}_{\text{Forces}} \qquad \underbrace{0 \quad 0 \quad 0}_{\text{Couples}}$$

Positions of suppressed displacements

1 2 3 4 5 6

Results

Displacements at node 1 are zero.
At free end (node 3), displacements are:

$$u_3^\circ = 0,\ v_3^\circ = 7.942\text{E-}3 \text{ m},\ w_3^\circ = 2.032\text{E-}3 \text{ m},$$
$$\theta_{x3}^\circ = 0,\ \theta_{y3}^\circ = -2.299\text{E-}3 \text{ rads},\ \theta_{z3}^\circ = 9.57\text{E-}3 \text{ rads}.$$

Forces and moments in units of N and m

Member	Node	Axial Force	Torque	BM_{x-y}	BM_{x-z}
1-2-4	1	0	0	−1000	2000
1-2-4	2	0	0	− 500	1000
2-3-4	2	0	0	− 500	1000
2-3-4	3	0	0	0	0

10

Bending moments in grillages

This chapter describes a computer program, (Appendix 10), which is suitable for the analysis of flat, horizontal cross-stiffened grids under vertical loads. The grids can be skew or orthogonal (Fig. 10.3), and their internal and external boundaries can be irregular.

The boundary conditions can be quite complex, so that various combinations of clamped and simply-supported edges can be catered for, together with discrete supports.

The vertical loading can be uniformly distributed in one direction with another value of uniformly distributed loading in the other direction. This loading can be in addition to concentrated vertical loads applied at the nodes.

Similarly, the sectional properties can have one value in one direction and another value in the other direction. If required, the program can be modified to allow for a much more comprehensive system of loading and sectional property distribution.

The method of analysis is the matrix displacement method, and the element in local co-ordinates is shown in Fig. 10.1.

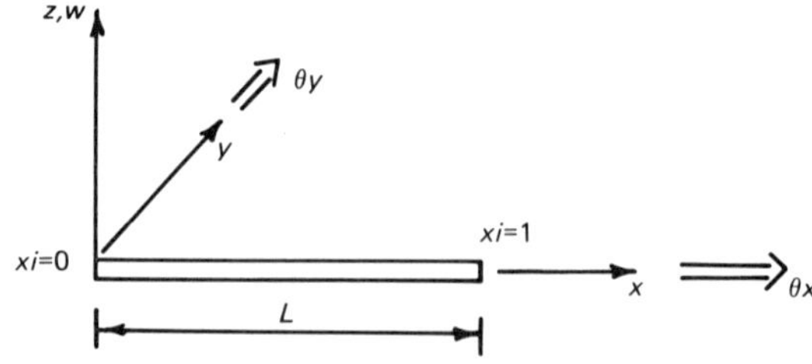

Fig. 10.1 – Beam element in local co-ordinates.

The element is also shown in Fig. 10.2 with respect to both local axes (x, y and z), and global axes (x°, y° and z°).

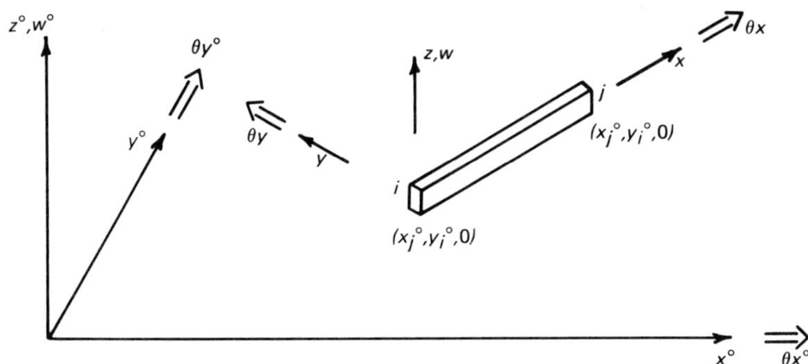

Fig. 10.2 – Beam element.

As the beam is a horizontal member, its local z axis is parallel to its global z° axis, so that $w = w^\circ$. The rotations $\theta_x{}^\circ$, $\theta_y{}^\circ$, etc. are according to the right-hand screw rule.

The elemental stiffness matrix and other particulars are given in [17], and will not be reproduced here. Solution of the simultaneous equations uses only half the band width of the stiffness matrix, thus, with a good choice of nodal numbering, it is possible to solve quite large grillages.

Fig. 10.3 shows the definition of X and Y direction members for an orthogonal grid and also for a skew grid.

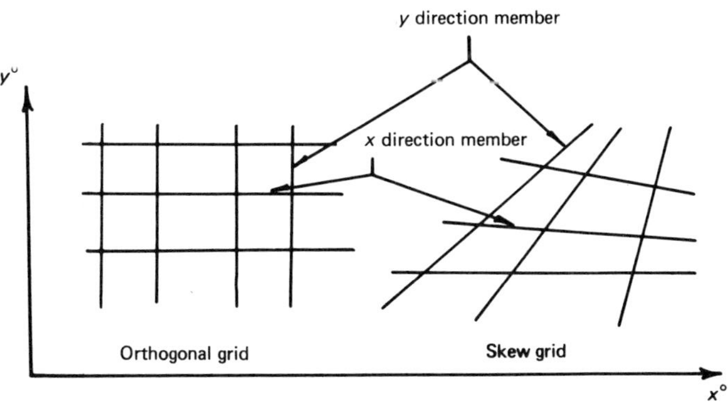

Fig. 10.3 – Orthogonal and Skew Grillages.

10.1 APPLICATIONS OF PROGRAMS

The data should be fed in as follows:

NJ = number of joints or nodes
NF = number of suppressed displacements
NC = number of concentrated vertical loads

Global co-ordinates of nodes

FOR $i = 1$ To NJ
→Input x_i°, y_i°

E = elastic modulus
G = rigidity modulus

AX = second moment of area of an X member about the $x^\circ - y^\circ$ plane
AY = second moment of area of a Y member about the $x^\circ - y^\circ$ plane
TX = torsional constant of an X member
TY = torsional constant of a Y member
MX = number of X members
MY = number of Y members
UX = load/unit length on an X member ($+^{ve}$ upwards)
UY = load/unit length on a Y member ($+^{ve}$ upwards)

Nodal points describing each member

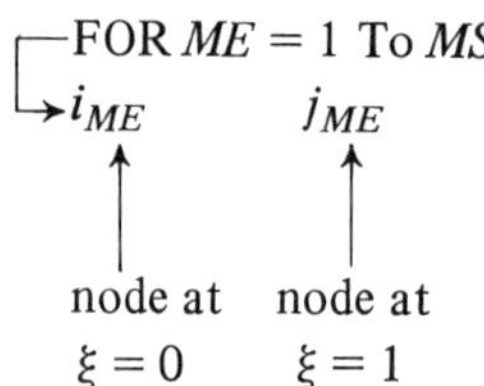

where $MS = MX + MY$; and the *nodes defining each of the X members must be fed in before feeding in the nodes defining any of the Y members*

Nodal positions and values of concentrated loads

FOR $i = 1$ To NC
Input (Nodal Position)$_i$
→Input (Vertical load)$_i$ – ($+^{ve}$ upwards)

Positions of Suppressed displacements

FOR $i = 1$ To NF
→Input NS_i – Displacements position of suppressed displacement

The positions of the *suppressed displacements* are calculated as follows:

At node i,
Suppressed displacement corresponding to $w_i^\circ = 3*i - 2$
Suppressed displacement corresponding to $\theta_{xi}^\circ = 3*i - 1$
Suppressed displacement corresponding to $\theta_{yi}^\circ = 3*i$

Example (on calculation of the suppressed displacement position)
If $NF = 6$, and the suppressed displacements are w_1°, θ_{x1}°, θ_{y2}°, w_9°, w_{15}° and θ_{x16}°, then the suppressed displacement positions are calculated as follows:

$$
\begin{aligned}
3*1-2 &= 1 \text{ (corresponding to } w_1^\circ) \\
3*1-1 &= 2 \text{ (corresponding to } \theta_{x1}^\circ) \\
3*2 &= 6 \text{ (corresponding to } \theta_{y2}^\circ) \\
3*9-2 &= 25 \text{ (corresponding to } w_9^\circ) \\
3*15-2 &= 43 \text{ (corresponding to } w_{15}^\circ) \\
3*16-1 &= 47 \text{ (corresponding to } \theta_{x16}^\circ)
\end{aligned}
$$

OUTPUT

The nodal displacements in global co-ordinates are output in ascending order, as follows:

$$w_1^\circ, \theta_{x1}^\circ, \theta_{y1}^\circ, w_2^\circ, \theta_{x2}^\circ \ldots w_{NJ}^\circ, \theta_{xNJ}^\circ, \theta_{yNJ}^\circ$$

Next, the torques and bending moments on individual elements are output in local co-ordinates.

Member	ξ	Torque	Bending moment
i–j	0	↓	↓
i–j	1	↓	↓
l–m	0	↓	↓
l–m	1	↓	↓

$\xi - 0$ is at thc 'origin' of the member (that is at $x = 0$)
$\xi = 1$ is at the other end of the member (that is at $x = L$)

Example 10.1

Determine the nodal displacements, torques and bending moments for the skew grillage shown in Fig. 10.4. The grillage may be assumed to be simply-supported at nodes 1, 4, 5, 8, 9 and 12, and the following apply:

$$
\begin{aligned}
AX &= 1.25\text{E} - 5 \text{ m}^4 & AY &= 1.25 \text{ E} - 5 \text{ m}^4 \\
TX &= 2.5 \text{ E} - 5 \text{ m}^4 & TY &= 2.5 \text{ E} - 5 \text{ m}^4 \\
E &= 2\text{E}11 \text{ N/m}^2 & G &= 7.69 \text{ E}10 \text{ N/m}^2 \\
UX &= -0.5 \text{ kN/m} & UY &= -1 \text{ kN/m}
\end{aligned}
$$

There is a downward load of 5 kN at node 7 (that is $W_7 = -5000$ N)

The data is as follows:

$NJ = 12$
$NF = 6$
$NC = 1$

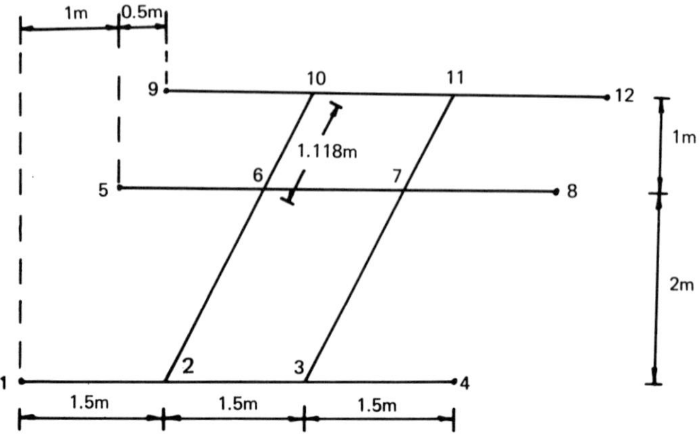

Fig. 10.4 – Skew grillage.

Nodal co-ordinates

$x_i°$	$y_i°$
0	0
1.5	0
3	0
4.5	0
1	2
2.5	2
4	2
5.5	2
1.5	3
3	3
4.5	3
6	3

$E = 2E11$
$G = 7.69E10$
$AX = 1.25E\text{-}5$
$AY = 1.25E\text{-}5$
$TX = 2.5E\text{-}5$
$TY = 2.5E\text{-}5$
$MX = 9$
$MY = 4$
$UX = -500$
$UY = -1000$

Nodal points defining each member

i	j
1	2
2	3
3	4
5	6
6	7
7	8
9	10
10	11
11	12
2	6
3	7
6	10
7	11

Nodal position and value of concentrated load

Nodal position = 7
Value of load = –5000

Displacement position of suppressed displacements

1 10 13 22 25 34

OUTPUT

Nodal displacements in global co-ordinates

Node (i)	w_i°(m)	θ_{xi}°	θ_{yi}°
1	0	– 2.11E-4	2.04E-3
2	– 2.58E-3	– 2.11E-4	1.11E-3
3	– 2.76E-3	– 9.87E-4	– 9.35E-4
4	0	– 9.87E-4	– 2.31E-3
5	0	6.91E-4	2.51E-3
6	– 3.13E-3	6.91E-4	1.27E-3
7	– 3.43E-3	– 1.26E-4	– 1.11E-3
8	0	– 1.26E-4	– 2.89E-3
9	0	8.4E-4	2.34E-3
10	– 2.86E-3	8.4E-4	1.07E-3
11	– 2.75E-3	3.2E-4	– 1.15E-3
12	0	3.21E-4	– 2.18E-3

Torques and bending moments in local co-ordinates

Member	ξ	Torque (Nm)	Bending Moment (Nm)
1-2	0	0	0
1-2	1	0	– 2910
2-3	0	– 995	– 2941
2-3	1	– 995	– 3684
3-4	0	0	– 4398
3-4	1	0	0
5-6	0	0	0
5-6	1	0	– 3933
6-7	0	– 1047	– 2659
6-7	1	– 1047	– 5108
7-8	0	0	– 5736
7-8	1	0	0
9-10	0	0	0
9-10	1	0	– 4063
10-11	0	– 664	– 3953
10-11	1	– 664	– 3261
11-12	0	0	– 3272
11-12	1	0	0
2-6	0	473	– 876
2-6	1	473	69.6
3-7	0	194	1209
3-7	1	194	– 2276
6-10	0	– 199	– 1437
6-10	1	– 199	643
7-11	0	287	– 1059
7-11	1	287	– 599

Solution of the above problem on a Commodore PET took under 7.5 minutes after the input of the data.

Example 10.2

Determine the nodal bending moments for the ship's grillage shown in Fig. 10.5, assuming the following to apply:

$AX = 8.6\text{E-}7 \text{ m}^4$ $\quad AY = 3.6\text{E-}6 \text{ m}^4$
$TX = 9\text{E-}8 \text{ m}^4$ $\quad TY = 2.2\text{E-}7 \text{ m}^4$
$E = 2\text{E}11 \text{ N/m}^2$ $\quad G = 7.69\text{E}10 \text{ N/m}^2$
$UX = -500 \text{ N/m}$ $\quad UY = 0$
$MX = 16$ $\quad MY = 14$

The grillage may be assumed to be simply-supported along its boundary, but there is no problem in allowing for clamped edges or discrete supports or a combination of complex boundary conditions. There are two downward loads of 4000 N and 3000 N at nodes 15 and 20 respectively.

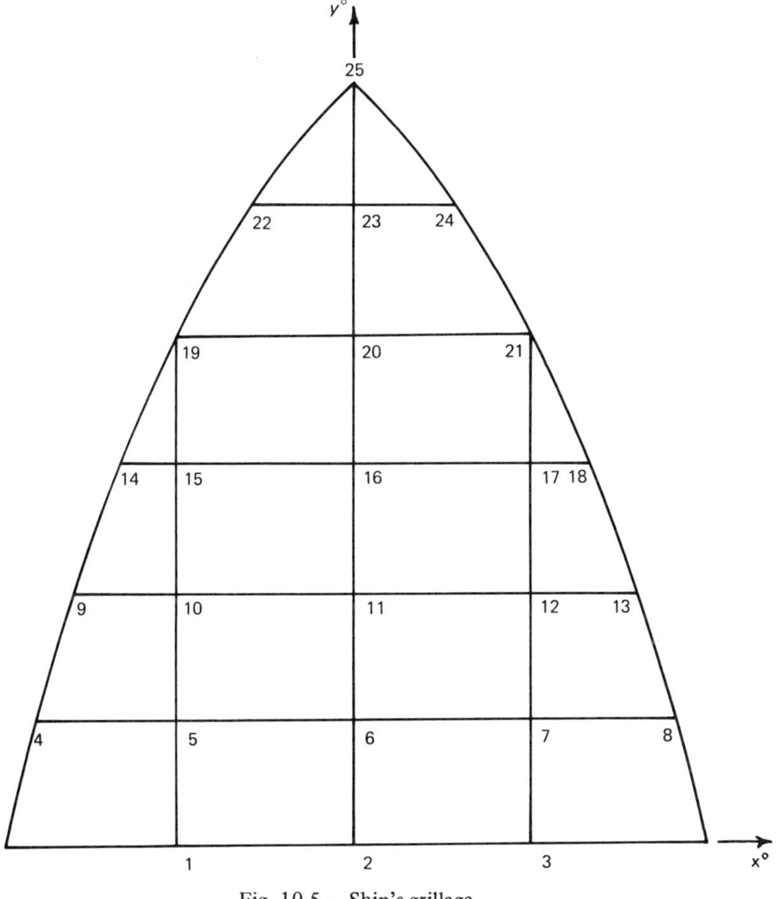

Fig. 10.5 – Ship's grillage.

Sketches of the bending moment distribution for some members are given in Fig. 10.6.

The time taken to solve the above problem on a CBM PET was under 19 minutes after the input of the data. The precision obtained was similar to that of a main frame computer.

In addition to Examples 10.1 and 10.2, the program successfully analysed Clarkson's grillage [17, 18] in about 23 mins – 15 s after the input of the data.

Prior to analysing Clarkson's grillage, the use of the instruction? FRE(0) revealed that there were 26595 bytes available on a 32 k PET. After running

Clarkson's grillage, use of the same instruction, revealed that there were 11422 bytes on the 32 k PET.

Solution of the 93 simultaneous equations, which had a half band width of 18, took about 11 minutes. Computation of an elemental stiffness matrix and its assembly took about 19.4 s/member, there being 38 members.

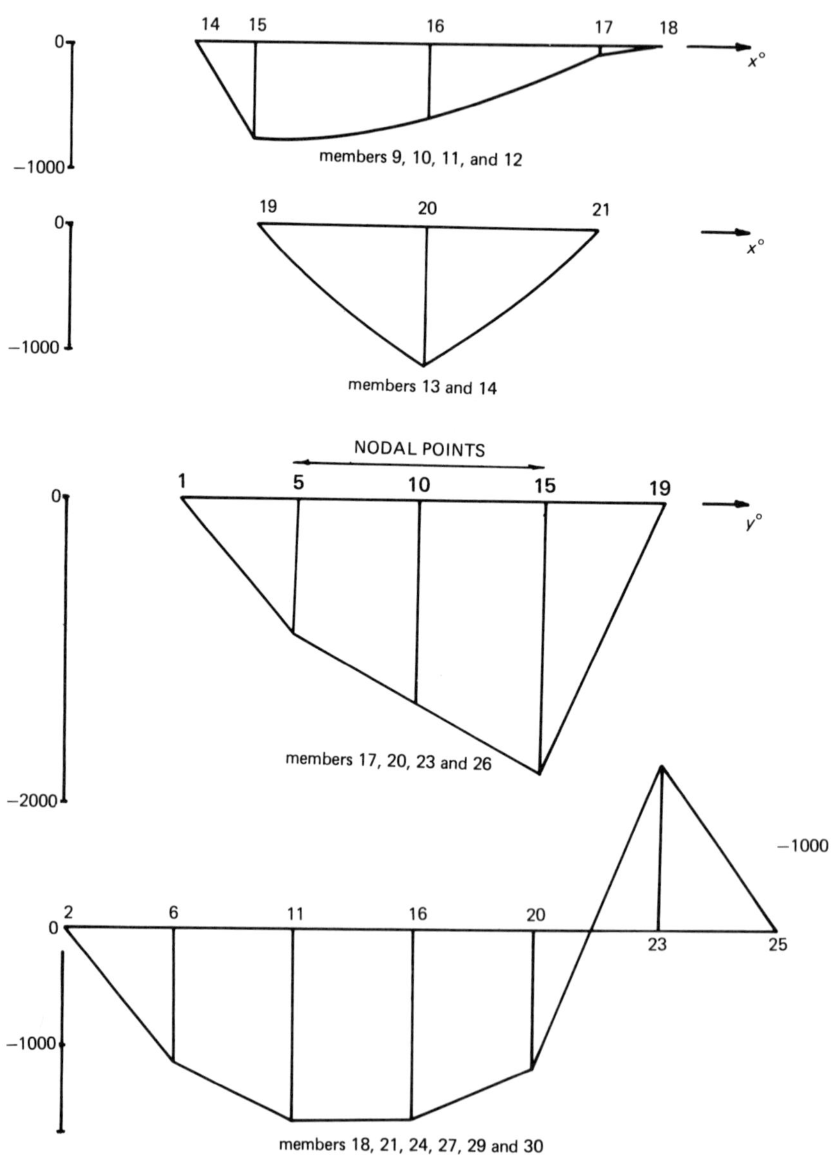

Fig. 10.6 – Bending moment distributions for ship's grillage.

11

Bending moments in bulkheads

This chapter describes a computer program (Appendix 11), which is suitable for the analysis of flat cross-stiffened grids under hydrostatic loading acting perpendicularly to the bulkhead surface, together with concentrated loads at the nodes. The grids can be skew or orthogonal (Fig. 10.3), and their internal and external boundaries can be irregular.

The boundary conditions can be quite complex, so that various combinations of clamped and simply-supported edges can be catered for, together with discrete supports. The sectional properties can have step variation in both directions, and the loading can be hydrostatic in both directions.

The method of analysis is the matrix displacement method, and the element in local co-ordinates is shown in Fig. 10.1.
The element is also shown in Fig. 10.2 with respect to both local axes (x, y and z), and global axes (x°, y° and z°).

The local z axis is parallel to the global z° axis so that $w = w^\circ$. The rotations $\theta_x{}^\circ, \theta_y{}^\circ$, etc., are according to the right-hand screw rule.

Solution of the simultaneous uses only half the band width of the stiffness matrix, thus, with a good choice of nodal numbering, it is possible to solve quite a large grillage.

Fig. 10.3 shows typical orthogonal and skew grillages.

11.1 APPLICATION OF PROGRAM

The data should be fed in as follows:

MS = number of elements or members
NJ = number of nodes
NF = number of suppressed displacements
NC = number of concentrated loads

Global co-ordinates of nodes

┌For $i = 1$ to NJ
│ Input x_i° y_i°
└→Next i

E = elastic modulus
G = rigidity modulus

Member details

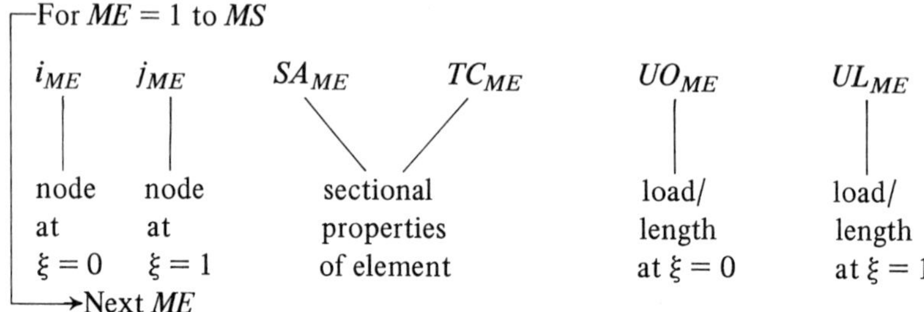

Nodal positions and values of concentrated loads

┌For $i = 1$ to NC
│ Input (Nodal Position)$_i$
│ Input (Concentrated Load)$_i$ ($+^{ve}$ in z° direction)
└→Next i

Positions of Suppressed Displacements

┌For $i = 1$ to NF
└→Input NS_i – Displacement position of suppressed displacement

The positions of the *suppressed displacements* are calculated as follows:
Suppressed displacement corresponding to $w_i^\circ = 3*i - 2$
Suppressed displacement corresponding to $\theta_{xi}^\circ = 3*i - 1$
Suppressed displacement corresponding to $\theta_{yi}^\circ = 3*i$

Example (on calculating the suppressed displacement positions)
If $NF = 6$, and the suppressed displacements are w_1°, θ_{x1}°, θ_{y2}°, w_9°, w_{15}° and θ_{x16}°, then the suppressed displacement positions are calculated as follows:

3*1-2 = 1 (corresponding to w_1°)
3*1-1 = 2 (corresponding to θ_{x1}°)
3*2 = 6 (corresponding to θ_{y2}°)
3*9-2 = 25 (corresponding to w_9°)
3*15-2 = 43 (corresponding to w_{15}°)
3*16-1 = 47 (corresponding to θ_{x16}°)

OUTPUT

The nodal displacements in global co-ordinates are output in ascending order, as follows:

$$w_1^\circ, \theta_{x1}^\circ, \theta_{y1}^\circ, w_2^\circ, \theta_{x2}^\circ \ldots, w_{NJ}^\circ, \theta_{xNJ}^\circ, \theta_{yNJ}^\circ$$

Next, the torques and bending moments on individual elements are output in local co-ordinates.

Member	ξ	Torque	Bending moment
i–j	0		
i–j	1		
l–m	0		
l–m	1		

$\xi = 0$ is at the 'origin' of the member (that is at $x = 0$)
$\xi = 1$ is at the other end of the member (that is at $x = L$)

Example 11.1

Determine the nodal displacements, torque and bending moments for the orthogonally stiffened bulkhead, hydrostatically loaded, as shown in Fig. 11.1. The bulkhead may be assumed to be clamped at nodes 1,2,3,4,5,6,12,18,19,20, 21 and 22, and simply-supported at nodes 11 and 17. For convenience, it will be assumed that the hydrostatic loading acts solely on the vertical stiffeners, and that this loading is transmitted to the horizontal girders at the joints that are common to both.

The following details may be assumed to apply to the bulkhead:

$$E = 2\text{E}8 \text{ kN/m}^2 \qquad G = 7.69\text{E}7 \text{ kN/m}^2$$

Stiffeners (Vertical members)

$$SA_{1-18} = SA_{2-19} = SA_{3-20} = SA_{4-21} = 8\text{E} - 4 \text{ m}^4$$
$$SA_{5-22} = 6\text{E} - 3 \text{ m}^2$$
$$TC_{1-18} = TC_{2-19} = TC_{3-20} = TC_{4-21} = 4\text{E} - 6 \text{ m}^4$$
$$TC_{5-22} = 3\text{E} - 4 \text{ m}^4$$

Girders (Horizontal members)

$$SA_{6-11} = 4\text{E} - 3 \text{ m}^4$$
$$SA_{12-17} = 2\text{E} - 3 \text{ m}^4$$
$$TC_{6-11} = 2\text{E} - 4 \text{ m}^4$$
$$TC_{12-17} = 1\text{E} - 4 \text{ m}^4$$

$$MS = 25$$

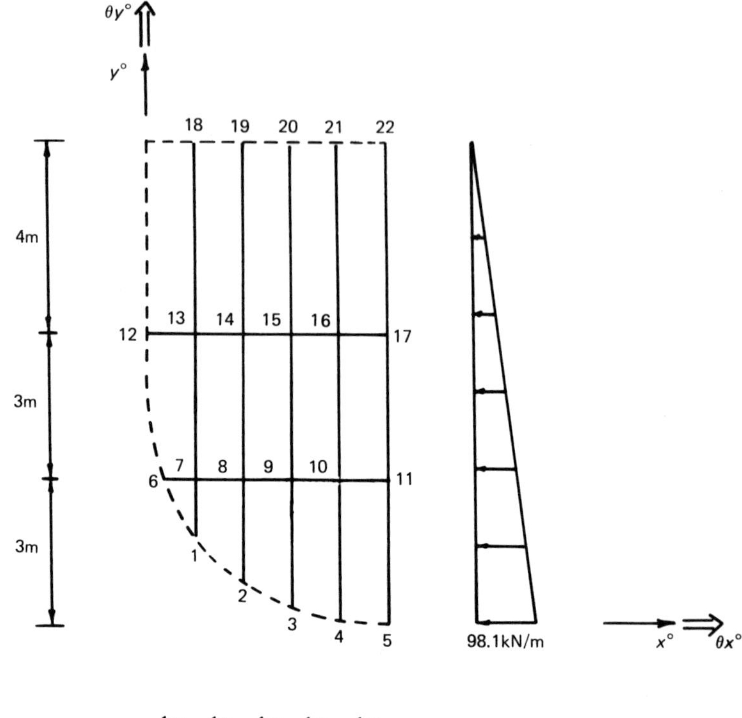

Fig. 11.1 – Hydrostatically loaded bulkhead.

The data is as follows:

$$MS = 25$$
$$NJ = 22$$
$$NF = 38$$
$$NC = 0$$

Nodal co-ordinates

x_i°	y_i°
1	1.75
2	0.8
3	0.25
4	0
5	0
0.3	3
1	3
2	3
3	3

4	3
5	3
0	6
1	6
2	6
3	6
4	6
5	6
1	10
2	10
3	10
4	10
5	10

$E = 2E8$
$G = 7.69E7$

Member details

i	*j*	*SA*	*TC*	*UO*	*UL*
6	7	4E-3	2E-4	0	0
7	8				
8	9				
9	10	↓	↓	↓	↓
10	11	4E-3	2E-4	0	0
12	13	2E-3	1E-4	0	0
13	14				
14	15				
15	16	↓	↓	↓	↓
16	17	2E-3	1E-4	0	0
1	7	8E-4	4E-6	80.9	68.7
2	8			90.3	
3	9	↓	↓	95.6	
4	10	8E-4	4E-6	98.1	↓
5	11	6E-3	3E-4	98.1	68.7
7	13	8E-4	4E-6	68.7	39.2
8	14				
9	15	↓	↓		
10	16	8E-4	4E-6	↓	↓
11	17	6E-3	3E-4	68.7	39.2
13	18	8E-4	4E-6	39.2	0
14	19				
15	20	↓	↓		
16	21	8E-4	4E-6	↓	↓
17	22	6E-3	3E-4	39.2	0

Displacement positions of suppressed displacements

1 2 3 4 5 6 7 8 9
10 11 12 13 14 15 16 17 18
31
34 35 36
49
52 53 54 55 56 57 58 59 60
61 62 63 64 65 66

OUTPUT

Moments and torques (kN.m) in local co-ordinates. (see Fig. 11.2)

Member	Torque	Moments ($\xi = 0$)	Moment ($\xi = 1$)
6-7	3.74	–457.5	–155.7
7-8	1.55	–155.8	142.1
8-9	–0.37	142.1	282.8
9-10	–1.79	282.8	240.0
10-11	–2.07	240.0	– 2.19
12-13	–0.48	–323.0	– 36.6
13-14	–0.47	– 36.7	128.5
14-15	–0.15	128.4	186.1
15-16	0.34	186.1	143.6
16-17	0.75	143.6	– 5.09
1-7	–0.066	– 33.5	– 29.5
2-8	–0.039	– 75.5	– 28.0
3-9	0	– 97.4	– 34.9
4-10	0.032	– 95.2	– 43.8
5-11	3.57	– 68.4	– 53.1
7-13	–0.019	– 31.7	– 24.3
8-14	0	– 30.0	– 8.9
9-15	0	– 36.3	– 3.1
10-16	0.016	– 44	– 12.4
11-17	1.38	– 51.1	– 32.3
13-18	0.035	– 24.3	– 33.0
14-19	0.026	– 8.5	– 53.6
15-20	0	– 2.6	– 61.1
16-21	–0.036	– 12.0	– 48.2
17-22	–3.71	– 33.1	– 20.0

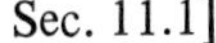

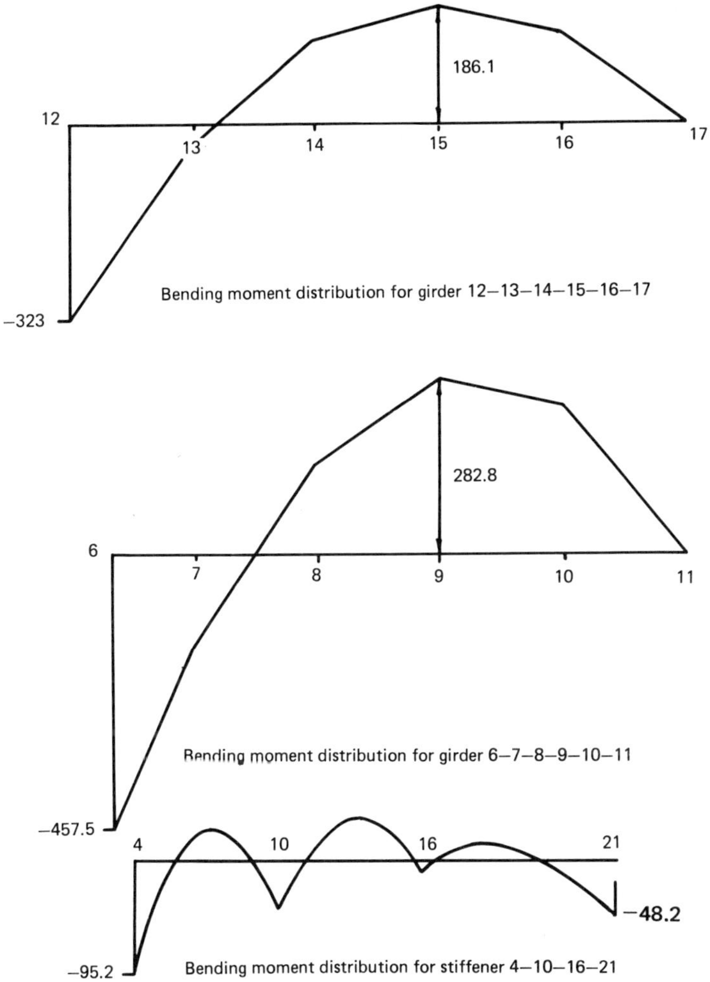

186.1

12 13 14 15 16 17

−323

Bending moment distribution for girder 12–13–14–15–16–17

282.8

6 7 8 9 10 11

−457.5

Bending moment distribution for girder 6–7–8–9–10–11

4 10 16 21

−48.2

−95.2

Bending moment distribution for stiffener 4–10–16–21

Fig. 11.2 – Bending moment distributions (kN/m).

After input of the data, the time taken to solve the above problem on a 32 k PET was 17 min. 25 s. The computation time taken for an element, up to its assembly, was about 18 s, there being 25 elements. The resulting 66 simultaneous equations were stored in a half band width of 21, and their solution was carried out in just under 10 minutes.

After solution of the problem on a 32 k PET, use of the instruction ?FRE(o), revealed that still larger problems can be tackled.

Analysis of the skew grillage of Example 10.1 by this program gave virtually the same results as that of the former.

NOTATION

With respect to the bulkhead problem, the following additional notation was used:

ξ	$= x/L$
SA	= 2nd moment of area of the cross-section of a member about the $x^\circ - y^\circ$ plane
TC	= torsional constant of the cross-section of a member
UO	= value of load/unit length as $\xi = 0$)
UL	= value of load/unit length as $\xi = 1$) see Fig. 11.3.

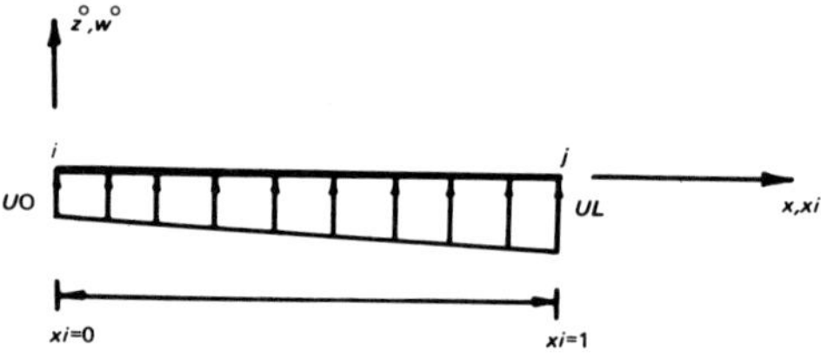

Fig. 11.3 – Member with hydrostatic load.

12

Stress analysis of thin-walled axisymmetric shells

The axisymmetric deformation of axisymmetric shells is of much importance in engineering and appears in a number of different forms, including submarine pressure hulls, containment vessels for nuclear reactors, condensers, storage vessels, etc.

For this program, (Appendix 12), the element used was similar to that of Grafton and Strome, [7, 19] but by employing numerical integration, it was possible to modify the element so that it could cater for the case where the shell thickness varied linearly along the meridian of the cone.

The lateral pressure could be uniform or hydrostatic and can be internal or external. It must be emphasised here, however, that if the lateral pressure is external, it is possible that failure may occur through instability and this feature should be considered when investigating vessels under external pressure. The pressure is assumed to act perpendicularly to the surface of the cone and the effects of axial pressure in addition to lateral pressure can be considered.

This element can be used for cylindrical or conical shells and can also be used for more complex axisymmetric shapes by representing the structure as a series of truncated cones, as shown in Fig. 12.1.

Solution of the resulting simultaneous equations involves only half the band width of the stiffness matrix, thus making a considerable saving in space and computational time.

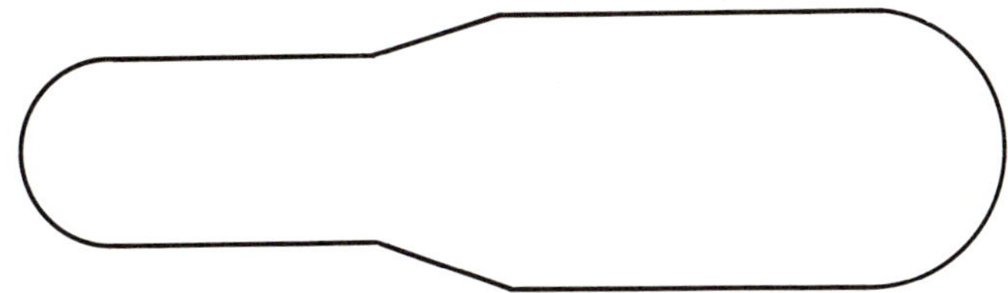

Fig. 12.1 – Complex Pressure Vessel.

12.1 ELEMENT DETAILS

The element used here is similar to that adopted by Grafton and Strome [19] the main difference being that as the element is tapered, integration is carried out numerically.

The element has two circumferential nodes at its extremities with three degrees of freedom per node, (u°, w° and θ), as shown in Fig. 12.2.

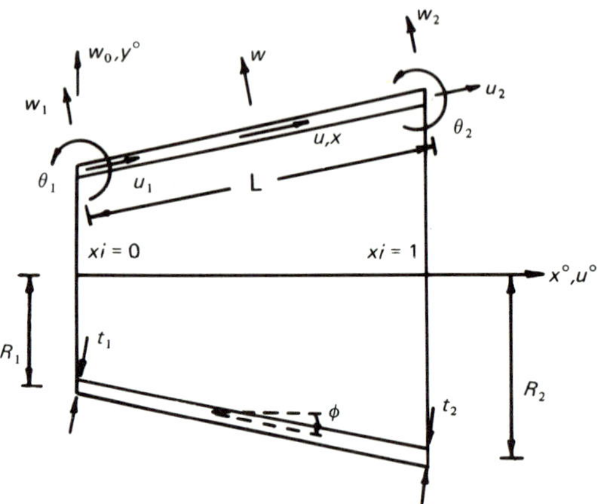

Fig. 12.2 – Thin-walled Conical Elements.

12.2 COMPUTER PROGRAM DETAILS

The program can determine axisymmetric displacements and stresses for thin-walled axisymmetric shells under axial and lateral pressure. The lateral pressure can be uniform, stepped or hydrostatic.

Similarly, the shell thickness can be uniform, stepped or linear and it is a simple matter to extend the program so that it can account for ring stiffeners.

One deficiency of the program is that in calculating stresses due to lateral pressure, the end fixing 'forces' have not been superimposed to give the net value of 'forces', but this effect can be minimised by ensuring that the elemental lengths are not too large. For a further discussion of this effect, see the discussion on distributed loads in [15, 20].

The data should be fed in as follows:

ES = number of elements

NF = number of suppressed displacements

E = elastic modulus

NU = Poisson's ratio

PX = axial pressure on closed end of vessel. (This is normally placed at the last node, and is positive if in the x° direction. Thus, if there is axial pressure, the last node must be free to move axially).

Displacement positions of suppressed displacements

$i = 1(1)NF$

$\rightarrow NS_i$ – the actual displacement position of each suppressed displacement.

Nodal details

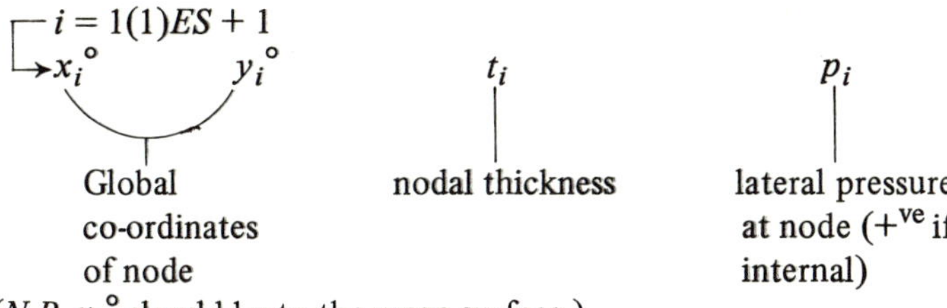

(*N.B.* y_i° should be to the mean surface.)

12.3 OUTPUT

$i = 1(1)NN$

$\rightarrow$ Node$_i$ u_i° w_i° θ_i – Global displacements of the nodes

where,

$$NN = ES + 1 = \text{number of nodes.}$$

Stresses

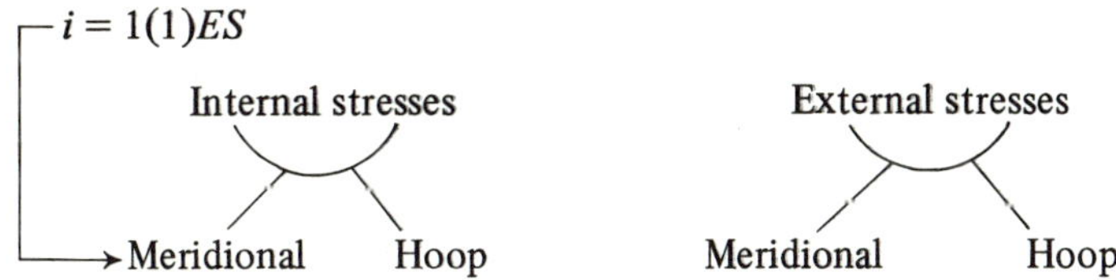

The stresses are determined at the two ends of the element, the first being at the origin of the element ($XI = 0$).

The following *shorthand notation* has been used for the input of vectors.

$i = 1(1)N$

$\rightarrow x_i\ y_i$

This means that the value of i is increased from one, in unit increments to a maximum value of N, then for each value of i, the corresponding values of x_i and y_i must be inputted.

Example 12.1

Determine axisymmetric displacements and stresses for the thin dome of [23]. The dome is in the form of a spherical shell cap of mean radius 90 in and wall thickness 3 in, as shown in Fig. 12.3. It is subjected to a uniform pressure of 1 lbf/in^2.

$$E = 3 \times 10^6 \text{ lbf/in}^2 \text{ (assumed)}$$

$$\upsilon = 1/6$$

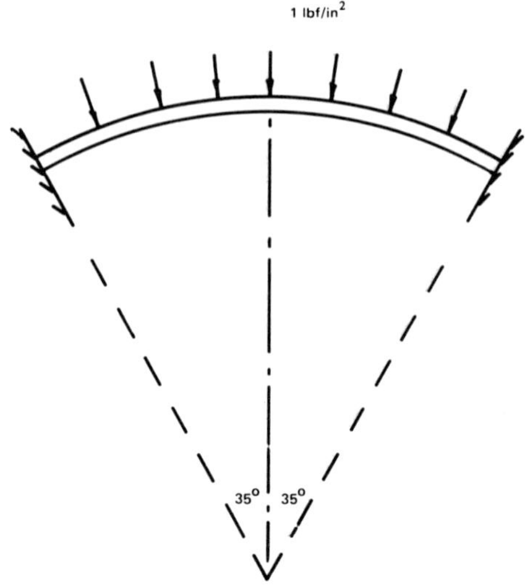

Fig. 12.3 – Spherical Shell Cap.

Comparison is made in Fig. 12.4 of the internal and external hoop forces/unit length with the average hoop force/unit length obtained by the analytical solution of Timoshenko and Woinowsky – Kreiger [23]. It can be seen that there is good agreement between the two solutions, as there is with the meridional bending moment/unit length of Fig. 12.5.

The data was as follows:

$$ES = 10$$
$$NF = 3$$
$$E = 3E6$$
$$NU = 0.16666667$$
$$PX = 0$$

Suppressed displacement positions

31 32 33

(These are all at node 11, and are calculated as follows:

Displacement position corresponding to $u_{11} \equiv 3 \times 11 - 2 = 31$
Displacement position corresponding to $w_{11} \equiv 3 \times 11 - 1 = 32$
Displacement position corresponding to $\theta_{11} \equiv 3 \times 11 = 33$)

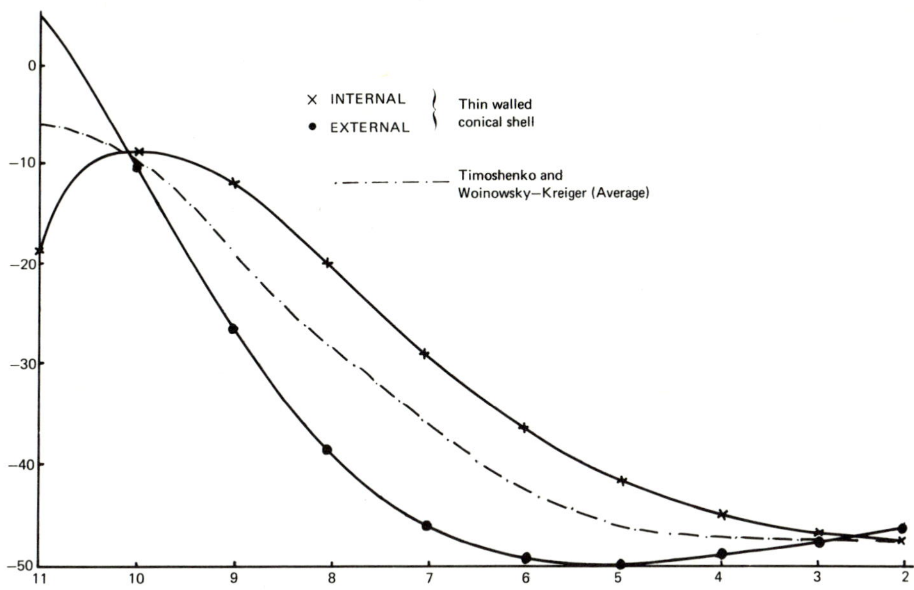

Fig. 12.4 – Hoop forces/unit length for cap.

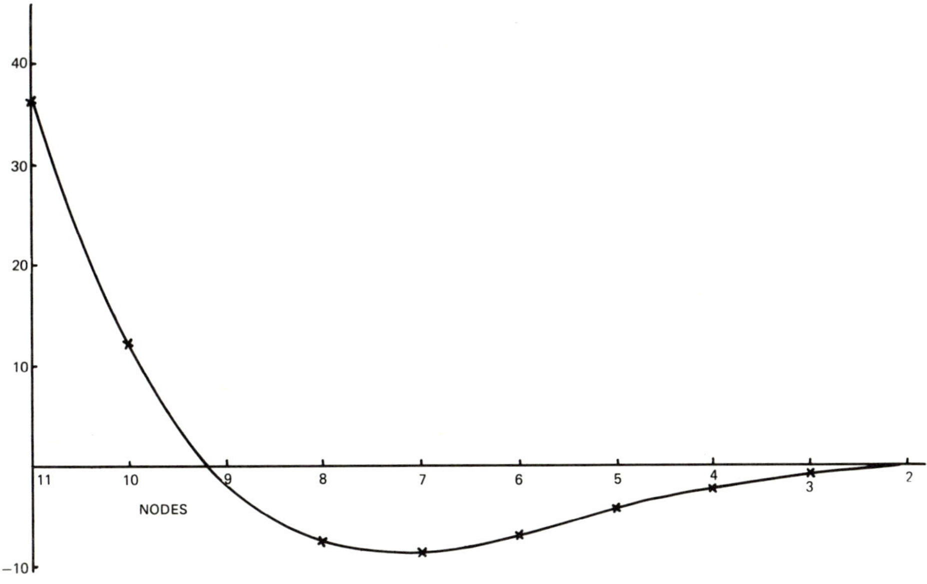

Fig. 12.5 – Meridional bending moment/unit length for cap.

Nodal details

x_i°	y_i°	t_i	p_i
0	0	3	–1
0.168	5.494	3	–1
0.671	10.968	3	–1
1.507	16.401	3	–1
2.673	21.773	3	–1
4.165	27.064	3	–1
5.978	32.253	3	–1
8.103	37.322	3	–1
10.535	42.252	3	–1
13.262	47.025	3	–1
16.277	51.622	3	–1

Results for Dome

Nodal displacements

Node	u_i°(in)	w_i°(in)	θ_i(rad)
1	5.47E–4	1.32E-9	–9.96E-8
2	5.46E–4	–2.41E-5	–1.40E-7
3	5.44E–4	–4.82E-5	2.49E-7
4	5.35E–4	–7.13E-5	1.62E-6
5	5.15E–4	–9.16E-5	4.41E-6
6	4.74E–4	–1.05E–4	8.86E-6
7	4.05E–4	–1.07E–4	1.47E-5
8	3.06E–4	–9.08E-5	2.07E-5
9	1.83E–4	–5.69E-5	2.390E–4
10	6.419E-5	–1.546E-5	1.942E-5
11	0	0	0

Some stress values (lbf/in^2)

Element	(xi)	Merid(int)	Hoop(int)	Merid(ext)	Hoop(ext)
1	1	–16.28	–16.01	–15.36	–15.63
2	0	–15.65	–15.91	–15.98	–15.73
2	1	–15.33	–15.63	–16.31	–16.00
– – –	– – –	– – –	– – –	– – –	– – –
9	0	–12.46	– 3.91	–14.96	– 8.74
9	1	–21.32	– 2.93	– 5.05	– 3.44
10	0	–20.68	– 2.88	– 4.66	– 3.32
10	1	–36.92	– 6.15	11.92	1.99

N.B.: $XI = 0$ is at origin end of element (that is, $x = 0$)
$XI = 1$ is at the other end of the element (that is, $x = 1$)

Example 12.2

Determine the axisymmetric displacements and stresses for Model number 7, [21], under a uniform circular cylinder of mean radius 5.165 in, wall thickness 0.08 in and is of length 10 in as shown in Fig. 12.6.

The vessel may be assumed to be clamped at both ends so that the ends are free to move axially towards each other.

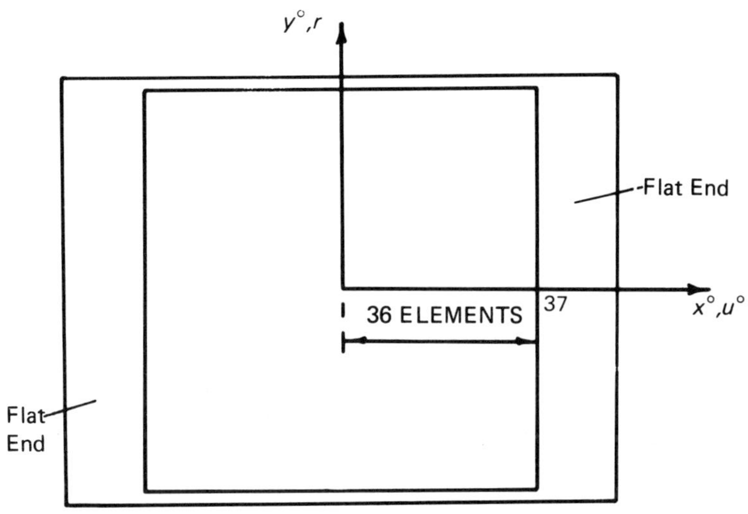

Fig. 12.6 – Thin-walled cylinder.

The property of symmetry was used about the centre of the vessel, and this half of the cylinder was divided into 36 elements as shown by the data. The first element was taken from the centre of the cylinder and the fineness of the mesh increased towards the clamped end.

The longitudinal and hoop stress distributions for the innermost fibre are shown in Fig. 12.7, where they are compared with the experimental observations.

Data for Model No. 7

$$ES = 36$$
$$NF = 4$$
$$E = 2.8E7$$
$$NU = 0.3$$
$$PX = -100$$

Suppressed displacement positions

1 3 110 111

Node	x_i°	y_i°	t_i	p_i
1	0	5.165	0.08	–100
2	.4			
3	.8			
4	1			
5	1.2			
6	1.4			
7	1.6			
8	1.8			
9	2.0			
10	2.2			
11	2.4			
12	2.6			
13	2.8			
14	3.0			
15	3.2			
16	3.4			
17	3.6			
18	3.7			
19	3.8			
20	3.9			
21	4.0			
22	4.1			
23	4.2			
24	4.3			
25	4.4			
26	4.5			
27	4.55			
28	4.6			
29	4.65			
30	4.7			
31	4.75			
32	4.8			
33	4.85			
34	4.9			
35	4.95			
36	4.975	↓	↓	↓
37	5.0	5.165	0.08	–100

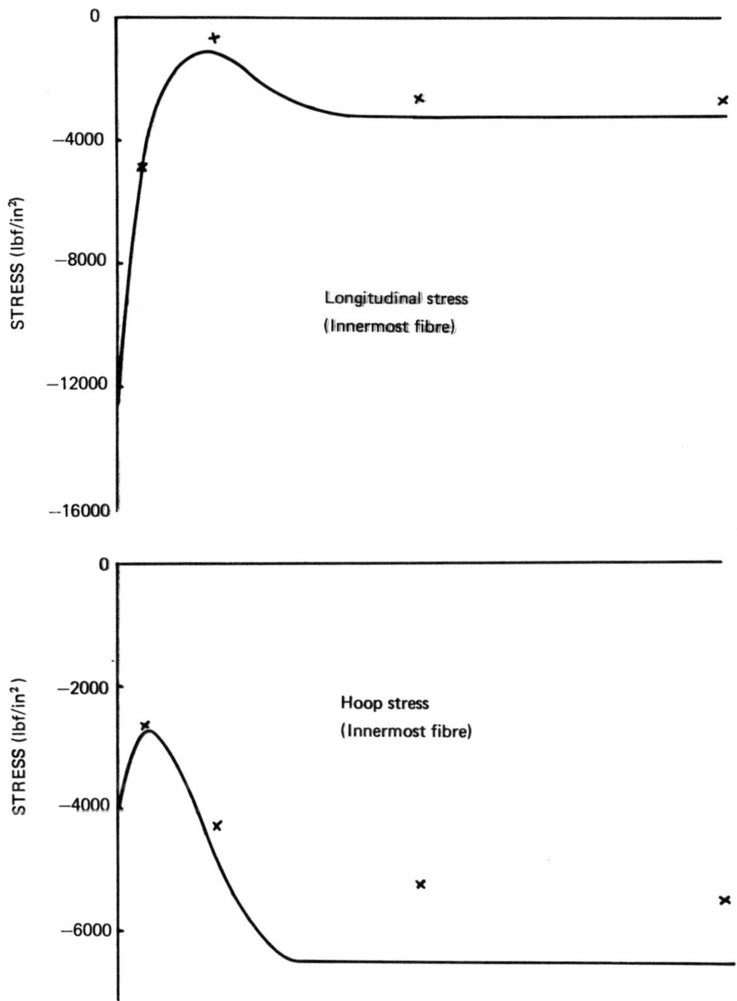

Fig. 12.7 – Longitudinal and Hoop Stress Distributions.

Results for Model No. 7

Some displacements

Node	u_i°(ins)	w_i°(ins)	θ_i(rads)
1	−4.26E-19	−1.01E-3	1.01E-18
2	−1.92E-5	−1.01E-3	1.72E-7
3	−3.83E-5	−1.01E-3	8.06E-7
– – –	– – –	– – –	– – –
37	−2.69E-4	0	0

Some stress values (lbf/in^2)

Element	(xi)	Merid(int)	Hoop(int)	Merid(ext)	Hoop(ext)
1	0	–3278	–6457	–3278	–6457
1	1	–3277	–6456	–3279	–6457
2	0	–3277	–6457	–3279	–6457
2	1	–3275	–6456	–3281	–6457
35	0	–11330	–3450	4760	1377
35	1	–12237	–3684	5692	1694
36	0	–12245	–3687	5683	1692
36	1	–13210	–3693	6656	1997

Example 12.3

Determine the axisymmetric displacements and stresses for the thin-walled cylinder shown in Fig. 12.8. This vessel has a stepped variation in wall thickness and is subjected to internal lateral hydrostatic pressure. It was first investigated in [12] by an analytical solution.

$$E = 30 \times 10^6 \text{ lbf/in}^2 \qquad \upsilon = 0.3$$

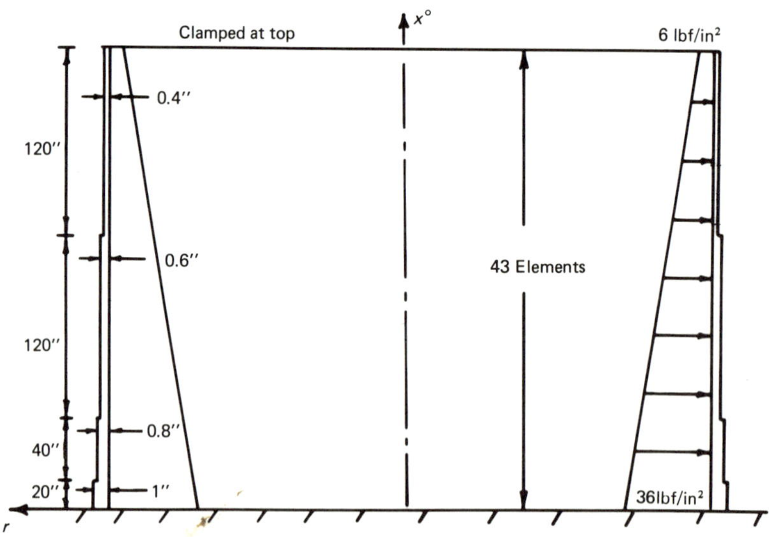

Fig. 12.8 – Vertical Tank.

The vessel was assumed to be clamped at the top end and fixed at the bottom end. It was divided into 43 elements, the mesh being finer at the two ends where the longitudinal bending moments were assumed to be large.

A plot is made of the computed radial deflection in Fig. 12.9. On comparison with [22], it was found that the finite element solution gave good results.

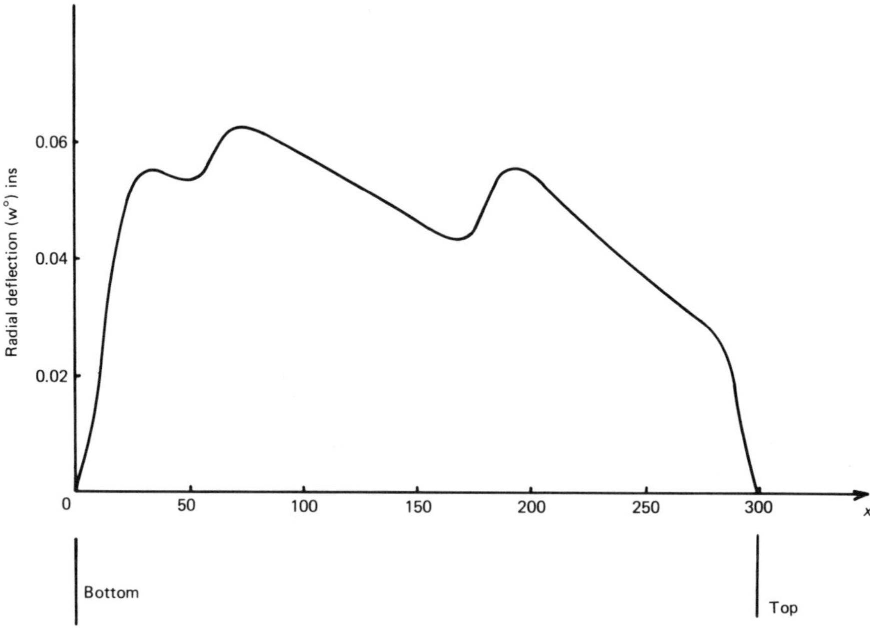

Fig. 12.9 – Radial displacement of vertical tank.

From the results obtained for the various examples considered, it can be seen that the thin-walled conical element is a very useful one and can be used for axisymmetric shells of various meridional shape.

The apparent discontinuities in forces and moments at some of the nodes is due to not superimposing the 'end fixing forces' due to pressure, but this effect can be minimised by taking smaller length elements. Further discussion of this effect is made in [15].

13

Plane stress and plane strain

This program, (Appendix 13), is intended for determining stresses in plates under in-plane forces. The program adopts the constant stress element of Clough [4], and it can be used for either plane stress or plane strain. The element is of triangular shape and it is described by nodes at its corners. Each node has 2 degrees of freedom, making a total of 6 degrees of freedom for the whole triangle, as shown in Fig. 13.1.

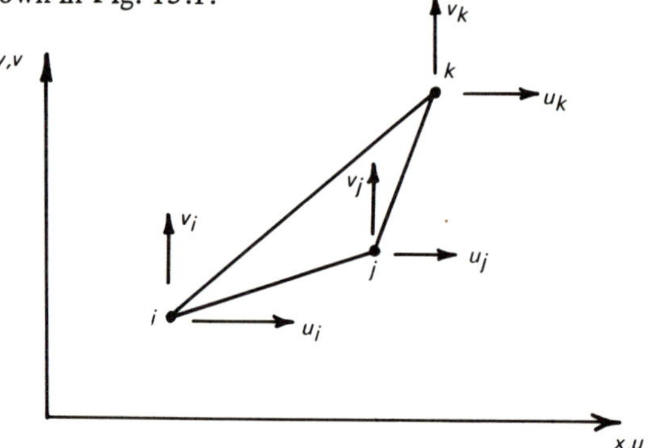

Fig. 13.1 – Triangular element with 3 nodes.

Direct stresses are calculated in the x and y directions, and shear stresses are calculated in the $x - y$ plane.

The stiffness matrix for the element is given in [4, 20, 24], although its derivation in the program is carried out numerically in global co-ordinates.

13.1 DATA

The data should be fed in as follows:

MS = number of elements
NN = number of nodes
NF = number of suppressed displacements
E = elastic modulus
NU = Poisson's ratio
T = plate thickness

If a plane stress analysis is required, feed 1, but if a plane strain analysis is required, feed zero.

Nodes describing each element

$ME = 1(1)MS$

$\rightarrow i_{ME}\ \ j_{ME}\ \ k_{ME}$ – anti-clockwise

Nodal co-ordinates

$i = 1(1)NN$

$\rightarrow x_i\ \ y_i$

Nodal positions and values of external loads

$i = 1(1)NC$

Nodal position of load

Horizontal component of load

$\rightarrow$ Vertical component of load

Suppressed displacement positions

$i = 1(1)NF$

$\rightarrow NS_i$

Example of suppressed displacement position calculation. Suppressed displacement position corresponding to

$$u_i \equiv 2 \times i - 1$$
$$v_i \equiv 2 \times i$$

That is, if in a plate, node 7 is completely suppressed, and node 9 is suppressed in the horizontal direction, then $NF = 3$, and,

Position 1 = 2 × 7 − 1 = 13)
) corresponding to node 7
Position 2 = 2 × 7 = 14)
Position 3 = 2 × 9 − 1 = 17 – corresponding to node 9

Example 13.1

Determine the nodal displacement and element stresses for the plate shown in Fig. 13.2 for (a) plane stress (b) plane strain. The plate, which is firmly fixed at nodes 1 and 2, is the same as that adopted as a worked example in [20, p. 32].

$$E = 6.93 \times 10^4\ \text{MN/m}^2,\ NU = 0.3,\ \text{Thickness} = 2 \times 10^{-2}\,\text{m}$$

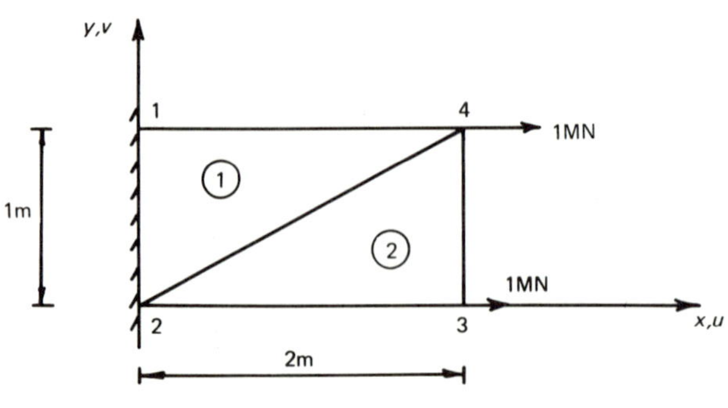

Fig. 13.2

The data is as follows:

$MS = 2$
$NN = 4$
$NF = 4$ – There are two suppressed displacements at node 1 and two at node 2.
$E = 6.93 \times 10^4$
$NU = 0.3$
$T = 2 \times 10^{-2}$
(a) $SS = 1$ or (b) $SS = 0$

Nodes describing each element

i	j	k
1	2	4
2	3	4

Nodal co-ordinates (global)

x_i	y_i		
0	1		
0	0		
2	0	2	1

$NC = 2$

Nodal positions and values of external loads

Nodal position 1 = 3
1 0

Nodal position 2 = 4
1 0

Suppressed displacement positions
1 2 3 4

RESULTS

(a) *Plane Stress*

$$u_1 = v_1 = u_2 = v_2 = 0$$

$u_4 = 2.639\text{E} - 3$ m $v_4 = 1.802\text{E} - 5$ m

$u_3 = 2.873\text{E} - 3$ m $v_3 = 4.506\text{E} - 4$ m

Element 1-2-4; $\sigma_x = 100.5$ MN/m^2 $\sigma_y = 30.1$ MN/m^2 $\tau_{xy} = 0.24$ MN/m^2

Element 2-3-4; $\sigma_x = 99.5$ MN/m^2 $\sigma_y = -0.12$ MN/m^2 $\tau_{xy} = -0.24$ MN/m^2

The results can be seen to agree favourably with the values given in [20].

(b) *Plane Strain*

$$u_1 = v_1 = u_2 = v_2 = 0$$

$u_3 = 2.568\text{E} - 3$ m $v_3 = 6.498\text{E} - 4$ m

$u_4 = 2.197\text{E} - 3$ m $v_4 = 9.28\text{E} - 5$ m

Element 1-2-4; $\sigma_x = 102.5$ MN/m^2 $\sigma_y = 43.92$ MN/m^2 $\tau_{xy} = 1.24$ MN/m^2

Element 2-3-4; $\sigma_x = 97.53$ MN/m^2 $\sigma_y = -0.62$ MN/m^2 $\tau_{xy} = -1.24$ MN/m^2

Example 13.2

Determine the nodal displacements and element stresses for the plate shown in Fig. 13.3 for plane stress. This example is the same as that adopted in [24, p. 26]. It has the following details:

$$E = 30 \times 10^6 \text{lbf/in}^2 \quad NU = 0.3$$

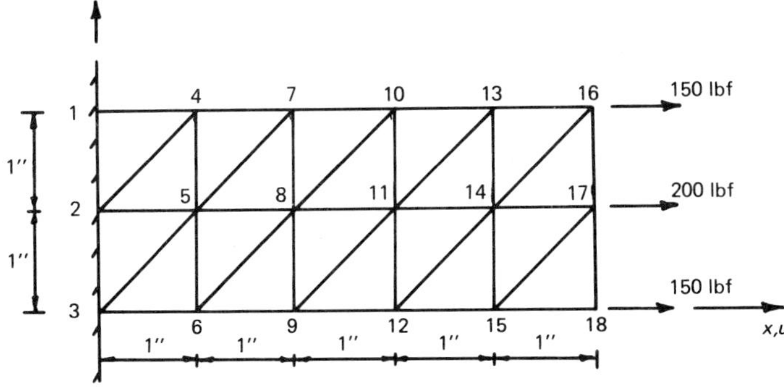

Fig. 13.3.

$MS = 20$
$NN = 18$
$NF = 6$ – There are two suppressed displacements at each of the nodes 1, 2 and 3.

$E = 3E7$
$NU = 0.3$
$T = 0.1$
$SS = 1$

Nodes describing each element

i	j	k
1	2	4
4	2	5
5	2	3
5	3	6
7	4	5
7	5	8
8	5	6
8	6	9
10	7	8
10	8	11
11	8	9
11	9	12
13	10	11
13	11	14
14	11	12
14	12	15
16	13	14
16	14	17
17	14	15
17	15	18

Nodal co-ordinates (global)

x	y
0	2
0	1
0	0
1	2
1	1
1	0
2	2
2	1

x	y
2	0
3	2
3	1
3	0
4	2
4	1
4	0
5	2
5	1
5	0

$NC = 3$

Nodal positions and values of concentrated loads

16	150	0
17	200	0
18	150	0

Suppressed displacement positions

1 2 3 4 5 6

RESULTS

Some nodal displacements

$$u_1 = v_1 = u_2 = v_2 = u_3 = v_3 = 0$$

$u_{16} = 4.164\text{E-}4$ $\quad v_{16} = 3.703\text{E-}6$
$u_{17} = 3.989\text{E-}4$ $\quad v_{17} = 2.821\text{E-}5$
$u_{18} = 4.335\text{E-}4$ $\quad v_{18} = 6.30\text{E-}5$

Some Stress Values

Element	σ_x(lbf/in^2)	σ_y(lbf/in^2)	τ_{xy}(lbf/in^2)
5-1-2	2464	739.2	–208
5-2-6	2344	87.9	–0.497
6-2-3	2548	764.3	28.77
6-3-7	2643	108.2	179.7
16-13-14	2749	69.5	69.91
16-14-17	2218	–69.9	250.6
17-14-15	2240	3.25	–113.9
17-15-18	2793	–206.6	–206.6

These results can be seen to compare favourably with the values given in [24].

Example 13.3

The square plate shown in Fig. 13.4 is firmly fixed at nodes 13, 14, 15 and 16. It has a square hole in its centre, and its particulars are as follows:

$$E = 1 \times 10^{11}\,\text{N/m}^2 \quad NU = 0.32 \quad \text{Thickness} = 2 \times 10^{-2}\,\text{m}$$

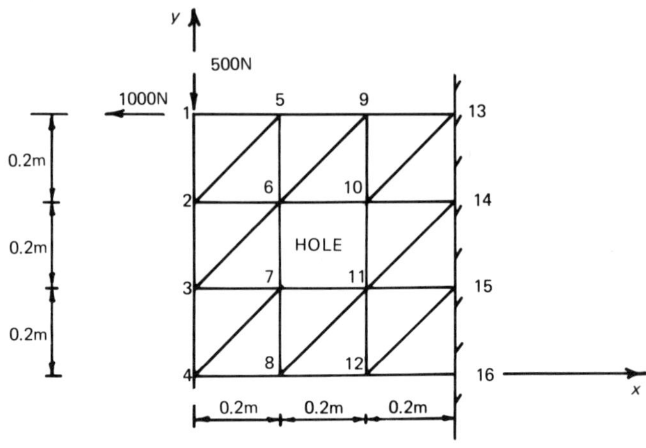

Fig. 13.4.

$MS = 16$
$NN = 16$
$NF = 8$
$E = 1E11$
$NU = 0.32$
$T = 2E\text{-}2$
$SS = 1$

i	j	k
5	1	2
5	2	6
6	2	3
6	3	7
7	3	4
7	4	8
9	5	6
9	6	10
11	7	8
11	8	12
13	9	10

i	j	k
13	10	14
14	10	11
14	11	15
15	11	12
15	12	16

x	y
0	0.6
0	0.4
0	0.2
0	0.0
0.2	0.6
0.2	0.4
0.2	0.2
0.2	0.0
0.4	0.6
0.4	0.4
0.4	0.2
0.4	0.0
0.6	0.6
0.6	0.4
0.6	0.2
0.6	0.0

$NC = 1$

1 –1000 –500

25 26 27 28 29 30 31 32

RESULTS

Some nodal displacements

$$u_{13} = v_{13} = u_{14} = v_{14} = u_{15} = v_{15} = u_{16} = v_{16} = 0$$

$u_1 = -3.026\text{E-}6$	$v_1 = -3.87\text{E-}6$
$u_2 = -8.29\text{E-}7$	$v_2 = -3.01\text{E-}6$
$u_3 = 1.89\text{E-}8$	$v_3 = -2.66\text{E-}6$
$u_4 = 8.38\text{E-}7$	$v_4 = -2.59\text{E-}6$

Some Stress Values

Element	σ_x(N/m^2)	σ_y(N/m^2)	τ_{xy}(N/m^2)
5-1-2	472,207	−277,793	−27,793
5-2-6	79,205	− 58,373	16,918
6-2-3	45,377	−164,086	96,789
6-3-7	− 24,902	− 26,063	79,339
13-9-10	517,537	43,848	−31,438
13-10-14	102,820	32,902	78,202
14-10-11	103,318	34,459	47,546
14-11-15	12,656	4,050	78,731
15-11-12	24,359	40,622	14,207
15-12-16	260,690	− 83,421	91,166

N.B. It should be noted that the mesh of Fig. 13.4 was very crude, and normally, a much more refined mesh would have been adopted.

14

Two dimensional field problems

This program, (Appendix 14), is intended for the solution of two dimensional field problems, of the Laplace and Poisson types.

It adopts a simplex triangular element with corner nodes, with one degree of freedom per node, as shown in Fig. 14.1, that is, the value of the unknown ϕ should be plotted perpendicularly to the plane of the triangle.

A linear distribution was assumed for ϕ over a triangular element, so that, the volume under any element $= (\phi_i + \phi_j + \phi_k)A/3$, where A = area of triangle.

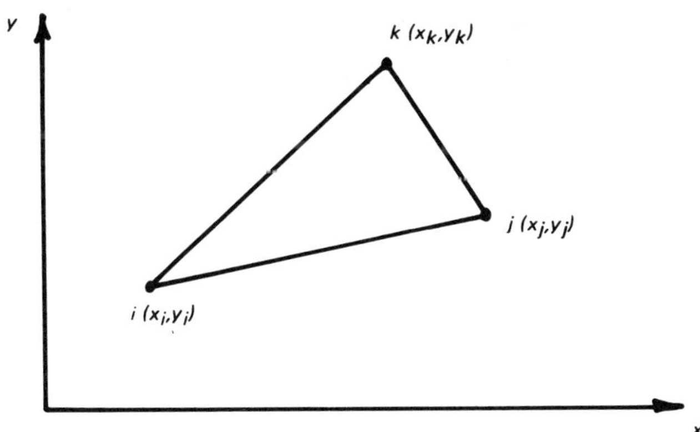

Fig. 14.1 – Triangular element.

The program calculates nodal values of ϕ and, (if required), slopes of ϕ in both x and y directions at the centroid of each element.

14.1 THEORY

The steady-state two-dimensional quasi-harmonic differential equation is as follows:

$$K\left(\frac{\partial^2 \phi}{\partial x^2} + \frac{\partial^2 \phi}{\partial y^2}\right) + Q = 0 \tag{14.1}$$

with the boundary conditions

(a) $\phi = \phi_B$ on the boundary

(b) and/or $K\left(l_x \dfrac{\partial \phi}{\partial x} + l_y \dfrac{\partial \phi}{\partial y}\right) + q_G + h_L\,(\phi - \phi_\infty) = 0$

where,

l_x and l_y are the direction cosines of a vector that is normal to the surface.

For the torsion of non-circular sections, (14.1) becomes

$$\frac{\partial^2 \phi}{\partial x^2} + \frac{\partial^2 \phi}{\partial y^2} + 2 = 0$$

with the boundary conditions,

(a) $\phi = \phi_B$ on the boundary

(b) and/or $\dfrac{\partial \phi}{\partial n} = l_x \dfrac{\partial \phi}{\partial x} + l_y \dfrac{\partial \phi}{\partial y} = 0$

The condition (b) is naturally achieved on a 'free' boundary, where,

ϕ = shear stress function.

From torsion theory, the following apply:

$$\begin{aligned} J &= \text{torsional constant} \\ &= 2 \times \text{volume under } \phi \\ &= 2 \int_{\text{Area}} \phi \, \mathrm{d}(A) \end{aligned}$$

Now,

$$\begin{aligned} T &= G\theta J \\ &= 2G\theta \int_{\text{Area}} \phi \, \mathrm{d}(A) \end{aligned}$$

where,

θ = angle of twist/unit length.

The shearing stresses are given by:

$$\left.\begin{aligned} \tau_{xz} &= G\theta \frac{\partial \phi}{\partial y} \\ \tau_{yz} &= -G\theta \frac{\partial \phi}{\partial x} \end{aligned}\right\} \text{ see Fig. 14.2.}$$

$$\tau = \sqrt{\tau_{xz}^2 + \tau_{yz}^2} = \text{resultant shear stress}$$

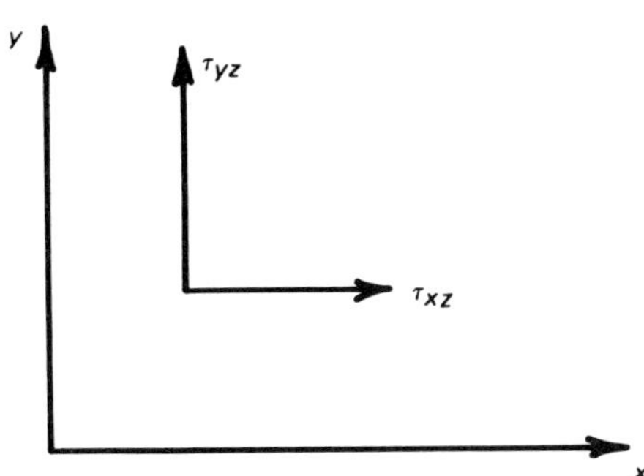

Fig. 14.2 – Shear stresses.

Heat Transfer

For steady-state heat transfer, (14.1) becomes

$$K\left(\frac{\partial^2 T}{\partial x^2}+\frac{\partial^2 T}{\partial y^2}\right)+Q=0$$

with boundary conditions,

(a) $T = T_B$, where T_B = boundary temperature

(b) and/or $K\left(l_x \frac{\partial T}{\partial x}+l_y \frac{\partial T}{\partial y}\right)+h_L(T-T_\infty)+q_G=0$

where,

T = Temperature (°K)
T_∞ = Known fluid temperature (°K)
K = Conductivity (kW/m °K)
Q = Heat generated within the body (kW/m^3) – $+^{ve}$ if heat is put into the body
h_L = Convection coefficient ($kW/m^2\,°K$)
q_G = Heat flux (kW/m^2) ($+^{ve}$ when heat is moving out of the body).

N.B. The heat flux q_G and the convection loss $h(T - T_\infty)$ must not take place over the same boundary.

Stream Functions

For the irrotational flow of an ideal fluid, (14.1) becomes,

$$\frac{\partial^2 \psi}{\partial x^2}+\frac{\partial^2 \psi}{\partial y^2}=0$$

where,

ψ = stream function

The flow velocities are obtained from

$$V_x = \frac{\partial \psi}{\partial y} \text{ and } V_y = -\frac{\partial \psi}{\partial x}$$

and the volume rate,

$$Q = \psi_j - \psi_i$$

where,

ψ_j and ψ_i are adjacent stream functions, and,
Q = flow rate/depth in z direction.

An alternative formulation

$$\frac{\partial^2 \phi}{\partial x^2} + \frac{\partial^2 \phi}{\partial y^2} = 0$$

$$V_x = \frac{\partial \phi}{\partial x} \text{ and } V_y = \frac{\partial \phi}{\partial y}$$

Also,

$$V_n = l_x \frac{\partial \phi}{\partial x} + l_y \frac{\partial \phi}{\partial y} = \frac{\partial \phi}{\partial n} = 0$$

where,

ϕ = velocity potential function.

It should be noted that the condition $\partial \phi / \partial n = 0$ occurs naturally over a 'free' boundary.

Ground water flow

For ground water flow in the horizontal $x - y$ plane of a confined aquifier, the constants in (14.1) have the following definitions:

K = coefficient of permeability (m^3/day/m^2)
ϕ = piezometric pressure head (m), measured from the bottom of the aquifier
Q = recharge (m^3/day) – when pumping, Q becomes $-^{ve}$.

The boundary conditions are:

(a) $\phi = \phi_B$ (boundary values of piezometric pressure head)

(b) and/or $K \frac{\partial \phi}{\partial n} + q = 0$

where,

q = seepage term representing water moving out of the boundary (m^3/day).

14.2 DATA

The data should be fed in as follows:

MS = number of elements
NN = number of nodes
NF = number of known nodal values of the function ϕ on the boundaries.

If **Q = constant** over the entire surface, **feed 1; else feed zero.**
If **Q = constant** over the entire surface, **feed Q.**
K = a constant (see appropriate equation).
If $\mathbf{h_L}$ **has a value** on any boundary, **feed 1; else feed zero.**
If $\mathbf{q_G}$ **has a value** on any boundary, **feed 1; else feed zero.**

Nodal points describing each element

┌ $ME = 1(1)MS$
└→ i_{ME} j_{ME} k_{ME} – anti-clockwise

Nodal co-ordinates

┌ $i = 1(1)NN$
└→ x_i y_i

Boundary conditions

┌ $i = 1(1)NF$
│ NS_i – nodal position of boundary
└→ ϕ_B– value of function at the above node

Other element details

$ME = 1(1)MS$

If Q is not constant over the entire surface, **feed Q** for this element.

If h_L or q_G have a value on any boundary for this element; **type Y**; else **type any other key**; unless both h_L and q_G are zero throughout.

Then feed,

either, LH = Number of boundaries on which h_L has a value

┌ $I = 1(1)LH$
│ i j – nodes defining boundary
│ h_L
└→ T_∞

and/or GQ = Number of boundaries on which q_G has a value

┌ $I = 1(1)GQ$
│ i j – nodes defining boundary
└→ q_G

Next ME

14.3 OUTPUT

$\phi_1 \quad \phi_2 \quad \phi_3 \qquad \phi_N$

$\dfrac{\partial\phi}{\partial x}$ and $\dfrac{\partial\phi}{\partial y}$ (optional) at element centroids

Example 14.1

Determine the torsional constant for the rectangular section shown in Fig. 14.3. The 8 element mesh should be used.

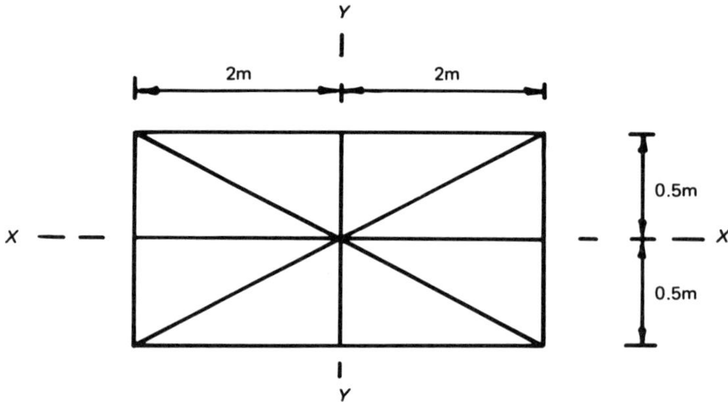

Fig. 14.3 – Rectangular section with 8 elements.

For this case, ϕ = zero on the boundaries, and the problem is symmetrical about the XX and YY axes. Hence, analysis can be carried out by considering only one quarter of the section, as shown in Fig. 14.4.

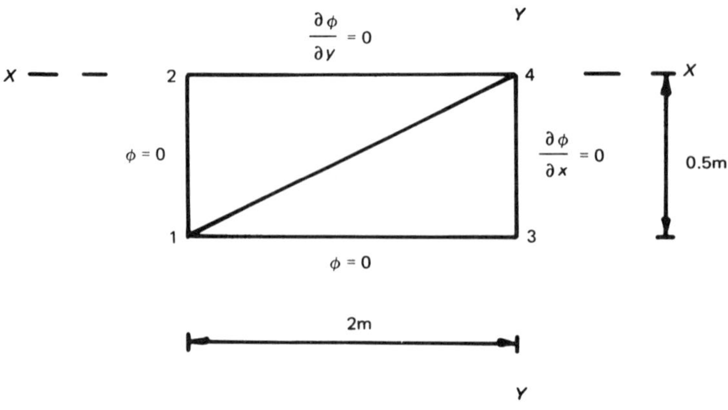

Fig. 14.4.

The data should be fed in as follows:

$$MS = 2$$
$$NN = 4$$
$$NF = 3$$

$$K = \begin{matrix} 1 & Q = 2 \\ 1 & \\ 0 & 0 \end{matrix}$$

i	j	k
4	2	1
4	1	3

Nodal co-ordinates

x_i	y_i
0	0
0	0.5
2	0
2	0.5

Boundary conditions

Nodes	Values of functions
1	0
2	0
3	0

Results

$$\phi_1 = \phi_2 = \phi_3 = 0$$

$$\phi_4 = 0.3137$$

Using numerical integration (that is, Simpson's rule),

$$J = 2 \times \text{volume under function}$$
$$\mathbf{J} = 1.115 \text{ m}^4$$

From 'exact' solution, $J = 1.12 \text{ m}^4$

Example 14.2

Determine the torsional constant for the square section shown in Fig. 14.5. The mesh shown should be adopted.

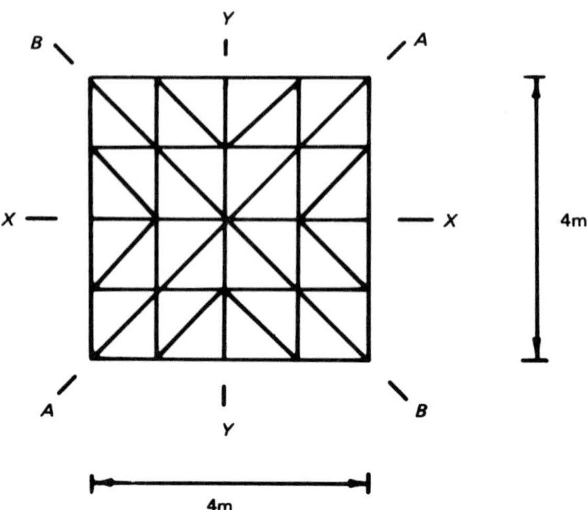

Fig. 14.5 – Square section with 32 elements.

For this problem, $\phi = 0$ on the boundaries and there are four axes of symmetry. Hence, solution can be carried out by considering only 1/8th of the section, as shown in Fig. 14.6.

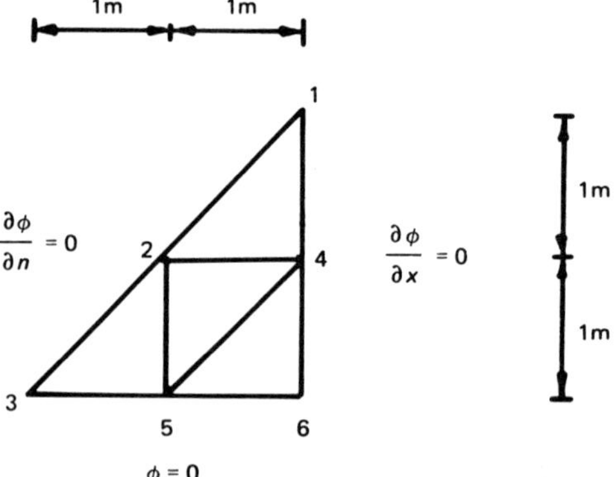

Fig. 14.6.

The data should be fed in as follows:

$MS = 4$

$NN = 6$

$NF = 3$

$$K = \begin{matrix} 1 & Q = 2 \\ 1 & \\ 0 & 0 \end{matrix}$$

i	j	k
1	2	4
2	3	5
4	2	5
4	5	6

Nodal co-ordinates

x_i	y_i
2	2
1	1
0	0
2	1
1	0
2	0

Boundary conditions

Node	Value of ϕ
3	0
5	0
6	0

Results

$$\phi_1 = 2.5 \quad \phi_2 = 1.417 \quad \phi_4 = 1.833$$
$$\phi_3 = \phi_5 = \phi_6 = 0$$

Using numerical integration, (that is, Simpson's rule),

$$J = 2 \times \text{volume under function}$$
$$\mathbf{J = 35.41\ m^4}$$

From 'exact' solution, $\mathbf{J = 36.05\ m^4}$

Example 14.3
Determine the values of the stream functions at the nodal points for the sudden enlargement problem shown in Fig. 14.7. It may be assumed that the problem is of unit thickness.

It will be convenient to assume that $\psi = 0$, along the bottom edge $4 \rightarrow 52$, and $\psi = 10$, along upper boundary $1 \rightarrow 19 \rightarrow 17 \rightarrow 47$.

The data should be fed in as follows:

$$MS = 74$$
$$NN = 52$$
$$NF = 22$$

```
      1    Q = 0
K = 1
      0    0
```

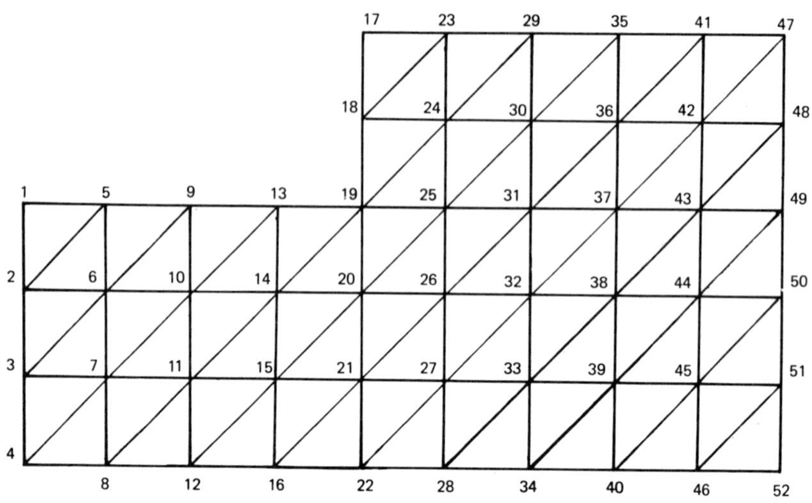

Fig. 14.7 – Sudden Enlargement.

i	j	k	i	j	k	i	j	k
5	1	2	23	18	24	38	32	33
5	2	6	24	18	19	38	33	39
6	2	3	24	19	25	39	33	34
6	3	7	25	19	20	39	34	40
7	3	4	25	20	26	41	35	36
7	4	8	26	20	21	41	36	42
9	5	6	26	21	27	42	36	37
9	6	10	27	21	22	42	37	43
10	6	7	27	22	28	43	37	38
10	7	11	29	23	24	43	38	44
11	7	8	29	24	30	44	38	39
11	8	12	30	24	25	44	39	45
13	9	10	30	25	31	45	39	40
13	10	14	31	25	26	45	40	46
14	10	11	31	26	32	47	41	42
14	11	15	32	26	27	47	42	48
15	11	12	32	27	33	48	42	43
15	12	16	33	27	28	48	43	49
19	13	14	33	28	34	49	43	44
19	14	20	35	29	30	49	44	50
20	14	15	35	30	36	50	44	45
20	15	21	36	30	31	50	45	51
21	15	16	36	31	37	51	45	46
21	16	22	37	31	32	51	46	52
23	17	18	37	32	38			

Nodal co-ordinates

x_i	y_i	x_i	y_i	x_i	y_i
0	6	8	6	14	6
0	4	8	4	14	4
0	2	8	2	14	2
0	0	8	0	14	0
2	6	10	10	16	10
2	4	10	8	16	8
2	2	10	6	16	6
2	0	10	4	16	4
4	6	10	2	16	2
4	4	10	0	16	0
4	2	12	10	18	10
4	0	12	8	18	8
6	6	12	6	18	6
6	4	12	4	18	4
6	2	12	2	18	2
6	0	12	0	18	0
8	10	14	10		
8	8	14	8		

Boundary conditions

4	0	8	0	12	0
16	0	22	0	28	0
34	0	40	0	46	0
52	0	1	10	5	10
9	10	13	10	19	10
18	10	17	10	23	10
29	10	35	10	41	10
47	10				

The results are as follows:

Node	ψ	Node	ψ
1	10	10	6.61
2	6.65	11	3.28
3	3.32	12	0
4	0	13	10
5	10	14	6.52
6	6.64	15	3.21
7	3.31	16	0
8	0	17	10
9	10	18	10

Node	ψ	Node	ψ
19	10	36	8.34
20	6.25	37	6.54
21	3.05	38	4.50
22	0	39	2.29
23	10	40	0
24	9.12	41	10
25	7.88	42	8.21
26	5.42	43	6.54
27	2.73	44	4.33
28	0	45	2.20
29	10	46	0
30	8.61	47	10
31	6.96	48	8.18
32	4.84	49	6.28
33	2.47	50	4.28
34	0	51	2.17
35	10	52	0

Example 14.4

Determine the temperatures at the nodal points for the heat transfer problem shown in Fig. 14.8. It will be assumed that the figure is of unit thickness.

$$K = 1000 \text{ W/(m °K)}$$
$$Q = 100 \text{ W/m}^3$$

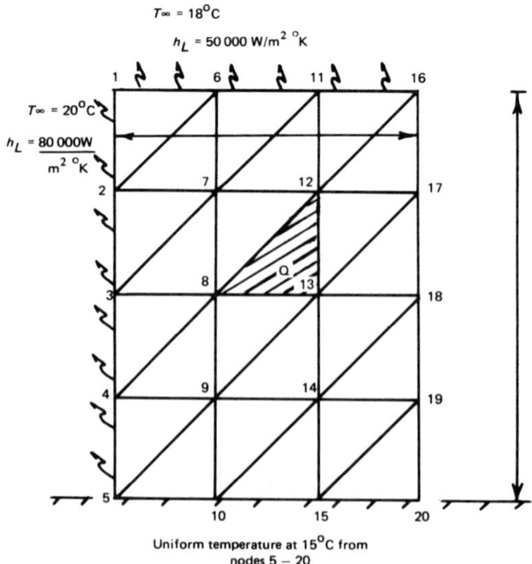

Fig. 14.8 – Heat transfer problem.

The data should be fed in as follows:

$MS = 24$
$NN = 20$
$NF = 4$
0
$K = 1000$
1 0

i	j	k	i	j	k	i	j	k
6	1	2	11	6	7	16	11	12
6	2	7	11	7	12	16	12	17
7	2	3	12	7	8	17	12	13
7	3	8	12	8	13	17	13	18
8	3	4	13	8	9	18	13	14
8	4	9	13	9	14	18	14	19
9	4	5	14	9	10	19	14	15
9	5	10	14	10	15	19	15	20

Nodal co-ordinates

x_i	y_i
0	8
0	6
0	4
0	2
0	0
2	8
2	6
2	4
2	2
2	0
4	8
4	6
4	4
4	2
4	0
6	8
6	6
6	4
6	2
6	0

5	15	10	15	15	15	
20	15					
0	Y	2	1	2	80,000	20
			1	6	50,000	18
0	N					
0	Y	1	2	3	80,000	20
0	N					
0	Y	1	3	4	80,000	20
0	N					
0	Y	1	4	5	80,000	20
0	N					
0	Y	1	6	11	50,000	18
0	N					
0	N					
100	N					
0	N					
0	N					
0	N					
0	N					
0	Y	1	11	16	50,000	18
0	N					
0	N					
0	N					
0	N					
0	N					
0	N					
0	N					

RESULTS

Node	T°C	Node	T°C
1	19.20	11	18.08
2	20.27	12	18.11
3	19.61	13	17.73
4	21.28	14	16.78
5	15.0	15	15.00
6	17.71	16	17.96
7	18.64	17	17.91
8	18.48	18	17.47
9	17.88	19	16.51
10	15.00	20	15.00

References

[1] Levy, S., Computation of Influence Coefficients for Aircraft Structures with Discontinuities and Sweepback, *J. Aero. Sci.,* **14**, 547–560, Oct. 1947.

[2] Levy, S., Structural Analysis and Influence Coefficients for Delta Wings, *J. Aero. Sci.,* **20**, 449–454, July, 1953.

[3] Argyris, J. H., Energy Theorems and Structural Analysis, *Aircraft Eng.,* Oct–Nov., 1954 and Feb–May, 1955.

[4] Turner, M. J., Clough, R. W., Martin, H. C. and Topp, L. J., Stiffness and Deflection Analysis of Complex Structures, *J. Aero. Sci.,* **23**, 805–823, 1956.

[5] Zienkiewicz, O. C. and Cheung, Y. K., Finite Element Analysis for Arch Dams and Comparison with Finite Difference Precedures, *Int., Symp., on Theory of Arch Dams, Southampton,* 123–140, 1964.

[6] Melosh, R. J., Basis for Derivation of Matrices for the Direct Stiffness Method, *J. A. I. A. A.,* **1**, 1631–1637, 1965.

[7] Zienkiewicz, O. C., *The Finite Element Method in Engineering Science,* McGraw-Hill, 1971.

[8] Richards, T. H., *Energy Methods in Stress Analysis,* Ellis Horwood, 1977.

[9] Hunter, S. C., *Mechanics of Continuous Media,* Ellis Horwood, 1976.

[10] Irons, B. and Ahmad, S., *Techniques of Finite Elements,* Ellis Horwood, 1980.

[11] Ross, C. T. F. and Johns, T., *Computer Analysis of Skeltal Structures,* Spon, 1981.

[12] Beards, C. F., *Vibration Analysis and Control System Dynamics,* Ellis Horwood, 1981.

[13] LaFara, R. L., *Computer Methods for Science and Engineering,* Hayden Book Co. Inc., 1973.

[14] Monro, D. M., *Interactive Computing with BASIC,* Arnold 1974.

[15] Ross, C. T. F., Structural Analysis by the Matrix Displacement Method, *DATA (AUEW),* Richmond, Surrey, 1970–1971.

[16] Ross, C. T. F., Algol Programs for Structural Analysis, *TASS (AUEW),* Richmond, Surrey, 1977–1978.

[17] Ross, C. T. F., ALGOL Programs for Cross-Stiffened Flat Plates and Grillages, *TASS (AUEW)*, Richmond, Surrey, 1979–1980.

[18] Clarkson, J., *et al.*, Data Sheets for the Elastic Design of Flat Grillages under Uniform Pressure, *European Shipbuilding*, No. 8, 1959.

[19] Grafton, P. E. and Strome, D. R., Analysis of Axisymmetric Shells by the Direct Stiffness Method, *J. A. I. A. A.*, **1**, pp. 2342–2347, 1963.

[20] Ross, C. T. F., Finite Element Theory in Structural Mechanics, *TASS (AUEW)*, Richmond, Surrey, 1974/1975.

[21] Ross, C. T. F., The Collapse of Ring-Reinforced Circular Cylinders Under Uniform External Pressure, *Trans. R. I. N. A.*, **107**, pp. 375–394, 1965.

[22] Ross, C. T. F., Axisymmetric Deformation of a Varying Thickness Cylinder under a Varying Lateral Pressure, *Trans. N.E.C.I.E.S.*, **87**, pp. 13–16, 1970.

[23] Timoshenko, S. and Woinowsky-Krieger, S., *Theory of Plates and Shells*, McGraw-Hill, 1959.

[24] Ross, C. T. F., Algol Programs for Plane Stress and Plane Strain, *TASS (AUEW)*, Richmond, Surrey, 1979–1980.

Appendices 1–15

APPENDIX 1

```
160 PRINT:PRINT:PRINT"FORCES IN PLANE PIN JOINTED TRUSSES":PRINT:PRINT
170 PRINT "NUMBER OF NODES"
180 INPUT NN
190 PRINT "NUMBER OF MEMBERS"
200 INPUT ES
210 N2=NN*2
220 PRINT "NUMBER OF SUPPRESSED DISPLTS"
230 INPUT NF
240 DIM SK(4,4),DC(4,4)
250 DIM U(N2),SG(4),SL(4),X(NN),Y(NN)
260 DIM XA(ES),XE(ES),XI(ES),XJ(ES)
270 DIM NS(NF)
280 PRINT "SUPPRESSED DISPLT POSNS"
290 FOR II=1 TO NF
300 PRINT "POSITION";II
310 INPUT NS(II)
320 NEXT II
380 PRINT "FEED IN NODAL COORDINATES"
390 FOR II=1 TO NN
400 PRINT "X COORD FOR NODE";II
410 INPUT X(II)
420 PRINT "Y COORD FOR NODE";II
430 INPUT Y(II)
440 NEXT II
445 MX=0
450 FOR EL=1 TO ES
460 PRINT "I NODE FOR MEMBER";EL
470 INPUT I
480 PRINT "J NODE FOR MEMBER";EL
490 INPUT J
500 PRINT "CROSS SECTNL AREA FOR MEM";EL
510 INPUT AR
520 PRINT "ELASTIC MODULUS FOR MEM";EL
530 INPUT E
540 XI(EL)=I
550 XJ(EL)=J
560 XA(EL)=AR
570 XE(EL)=E
571 IFABS(J-I)>MX THEN MX=ABS(J-I)
572 NEXT EL
573 NW=(MX+1)*2:NT=N2+NW
574 DIM A(NT,NW),Q(NT),C(NT)
575 FOR I=1 TO NT:FOR J=1 TO NW:A(I,J)=0
576 NEXTJ:Q(I)=0:NEXTI
577 FOR EL=1 TO ES
578 I=XI(EL):J=XJ(EL):AR=XA(EL):E=XE(EL)
579 PRINT:PRINT:PRINT"MEMBER";EL;"UNDER COMPUTATION":PRINT:PRINT
580 L=SQR((X(J)-X(I))↑2+(Y(J)-Y(I))↑2)
590 KS=(X(J)-X(I))/L
```

```
600 SN=(Y(J)-Y(I))/L
610 SK(1,1)=KS↑2
620 SK(3,3)=KS↑2
630 SK(2,1)=KS*SN
640 SK(1,2)=KS*SN
650 SK(3,4)=KS*SN
660 SK(4,3)=KS*SN
670 SK(1,3)=-KS↑2
680 SK(3,1)=-KS↑2
690 SK(1,4)=-KS*SN
700 SK(4,1)=-KS*SN
710 SK(2,3)=-KS*SN
720 SK(3,2)=-KS*SN
730 SK(2,2)=SN↑2
740 SK(4,4)=SN↑2
750 SK(2,4)=-SN↑2
760 SK(4,2)=-SN↑2
770 CN=AR*E/L
780 FOR II=1 TO 4
790 FOR JJ=1 TO 4
800 SK(II,JJ)=SK(II,JJ)*CN
810 NEXT JJ
820 NEXT II
830 I1=2*I-2
840 J1=2*J-2
845 FOR JJ=1 TO 2
850 IF JJ=1 THEN NR=I1
855 IF JJ=2 THEN NR=J1
860 FOR J9=1 TO 2
865 NR=NR+1:II=(JJ-1)*2+J9
870 FOR KK=1 TO 2
875 IF KK=1 THEN N9=I1
880 IF KK=2 THEN N9=J1
885 FOR K=1 TO 2
890 LL=(KK-1)*2+K
895 NK=N9+K+1-NR
900 IF NK<=0 THEN 910
905 A(NR,NK)=A(NR,NK)+SK(II,LL)
910 NEXT K
915 NEXT KK,J9,JJ
970 NEXT EL
980 FOR II=1 TO NF
990 N9=NS(II)
1000 A(N9,1)=A(N9,1)*1E12+1E12
1010 NEXT II
1020 PRINT
1030 PRINT "FEED VECTOR OF NODAL LOADS":PRINT
1040 FOR II=1 TO N2
1050 PRINT "POSITION";II
1060 INPUT Q(II)
1065 C(II)=Q(II)
1070 NEXT II
1090 N=N2
1095 PRINT:PRINT:PRINT"SOLUTION OF THE SIMULTANEOUS EQUATIONS IS NOW BEING CARR
IED OUT":PRINT:PRINT
1100 GOSUB 2940
1110 FOR II=1 TO N2
1120 U(II)=C(II)
1160 NEXT II
1170 PRINT
1180 PRINT "THE NODAL DISPLTS ARE"
1190 FOR II=1 TO N2
1200 PRINT U(II);
1210 NEXT II
1220 FOR II=1 TO 4
1230 FOR JJ=1 TO 4
1240 DC(II,JJ)=0.0
1250 NEXT JJ
1260 NEXT II
1270 PRINT:PRINT:PRINT"FOR SUCCESSIVE VALUES OF ";
1280 PRINT "NODAL NUMBERS OF MEMBERS AND FORCES, PRESS ANY KEY":PRINT
1290 FOR EL=1 TO ES
1300 AR=XA(EL)
1310 E=XE(EL)
1320 I=XI(EL)
```

```
 1330 J=XJ(EL)
 1340 L=SQR((X(J)-X(I))↑2+(Y(J)-Y(I))↑2)
 1350 KS=(X(J)-X(I))/L
 1360 SN=(Y(J)-Y(I))/L
 1370 DC(1,1)=KS
 1380 DC(2,2)=KS
 1390 DC(3,3)=KS
 1400 DC(4,4)=KS
 1410 DC(1,2)=SN
 1420 DC(3,4)=SN
 1430 DC(2,1)=-SN
 1440 DC(4,3)=-SN
 1450 I1=2*I-2
 1460 J1=2*J-2
 1470 FOR II=1 TO 2
 1480 MM=I1+II
 1490 NM=J1+II
 1500 SG(II)=U(MM)
 1510 SG(II+2)=U(NM)
 1520 NEXT II
 1530 FOR II=1 TO 4
 1540 SL(II)=0
 1550 FOR JJ=1 TO 4
 1560 SL(II)=SL(II)+DC(II,JJ)*SG(JJ)
 1570 NEXT JJ
 1580 NEXT II
 1590 FC=AR*E*(SL(3)-SL(1))/L
 1600 PRINT I;J;FC
 1610 GETA$:IFA$=""THEN GOTO1610
 1620 NEXT EL
 1630 END
 2940 FOR II=1 TO N
 2950 IK=II
 2960 FOR JJ=2 TO NW
 2970 IK=IK+1
 2980 CN=A(II,JJ)/A(II,1)
 2990 JK=0
 3000 FOR KK=JJ TO NW
 3010 JK=JK+1
 3020 A(IK,JK)=A(IK,JK)-CN*A(II,KK)
 3030 NEXT KK
 3040 A(II,JJ)=CN
 3050 C(IK)=C(IK)-CN*C(II)
 3060 NEXT JJ
 3070 C(II)=C(II)/A(II,1)
 3080 NEXT II
 3090 FORIZ=2TON
 3100 II=N-IZ+1
 3110 FORKK=2TONW
 3120 JJ=II+KK-1
 3130 C(II)=C(II)-A(II,KK)*C(JJ)
 3140 NEXTKK,IZ
 3150 RETURN
READY.
```

APPENDIX 2

```
 8 PRINT:PRINT:PRINT"BENDING MOMENTS IN CONTINUOUS BEAMS":PRINT:PRINT
 9 PRINT:PRINT:PRINT"THIS PROGRAM STORES ONLY THAT PART OF THE STIFFNESS MATRIX
CONTAINED";
 10 PRINT" WITHIN THE UPPER HALF OF ITS BANDWIDTH(NW)- N.B. NW=4 FOR THIS CASE":
PRINT:PRINT
 15 PRINT:PRINT:PRINT"IT DOES NOT INVERT A MATRIX, BUT SOLVES THE SIMULTANEOUS E
QUATIONS";
 16 PRINT" THROUGH TRIANGULATION":PRINT:PRINT
 20 PRINT "FEED IN THE NUMBER OF ELEMENTS"
 30 INPUT LS
 40 PRINT "NUMBER OF SUPPRESSED DISPLTS"
 45 INPUT NF
 50 NW=4
 55 NN=2*LS+2
 60 NT=NN+NW
 70 DIM A(NT,NW),Q(NT),C(NT),U(NN),ST(4,4),UD(LS),UP(LS),SA(LS),XL(LS)
 80 DIM DI(4),AS(LS),B(4),NS(NF)
 130 ND=LS+1
 140 FOR II=1 TO NT
 150 FOR JJ=1 TO NW
 160 A(II,JJ)=0.0
 170 NEXT JJ
 180 Q(II)=0.0
 190 NEXT II
 200 PRINT "NO OF CONCENTRATED LOADS"
 210 INPUT NC
 220 PRINT "POSN OF SUPPRESSED DISPLTS"
 230 FOR II=1 TO NF
 235 PRINT "POSITION";II
 240 INPUT NS(II)
 250 NEXT II
 260 PRINT "ELASTIC MODULUS"
 270 INPUT E
 280 PRINT "IF HYDROSTATIC LOAD FEED 1,ELSE FEED 0.0"
 290 INPUT NL
 300 FOR LE=1 TO LS
 350 PRINT "2ND MOM OF AREA FOR ELEM";LE
 360 INPUT SA(LE)
 370 PRINT "ELEMENTAL LENGTH FOR ELEM";LE
 380 INPUT XL(LE)
 390 IF NL=1 THEN 430
 400 PRINT "DISTRIBUTED LOAD FOR ELEM";LE
 410 INPUT UD(LE)
 420 GOTO 490
 430 PRINT "HYDRO LOAD AT A(ON LEFT) FOR ELEM";LE
 440 INPUT UD(LE)
 450 PRINT "HYDRO LOAD (AT RIGHT) FOR ELEM";LE
 460 INPUT UP(LE)
 470 PRINT "THE LENGTH A FOR ELEM";LE
 480 INPUT AS(LE)
 490 I=LE
 500 J=I+1
 510 L=XL(LE)
 520 IF NL=0 THEN SW=UD(LE)
 530 ST(1,1)=12/L↑3
 540 ST(1,2)=-6/L↑2
 550 ST(1,4)=ST(1,2)
 560 ST(2,1)=ST(1,2)
 570 ST(4,1)=ST(1,2)
 580 ST(3,3)=ST(1,1)
 590 ST(1,3)=-12/L↑3
 600 ST(3,1)=ST(1,3)
 610 ST(2,2)=4/L
 620 ST(4,4)=4/L
 630 ST(2,3)=6/L↑2
 640 ST(3,2)=ST(2,3)
 650 ST(2,4)=2/L
 660 ST(4,2)=2/L
 670 ST(3,4)=6/L↑2
 680 ST(4,3)=ST(3,4)
 690 CN=E*SA(LE)
 700 FOR II=1 TO 4
```

```
710 FOR JJ=1 TO 4
720 ST(II,JJ)=ST(II,JJ)*CN
730 NEXT JJ
740 NEXT II
750 I1=2*I-2
760 J1=2*J-2
770 IF NL=1 THEN 830
780 Q(I1+1)=Q(I1+1)+SW*L/2
790 Q(I1+2)=Q(I1+2)-SW*L↑2/12
800 Q(J1+1)=Q(J1+1)+SW*L/2
810 Q(J1+2)=Q(J1+2)+SW*L↑2/12
820 GOTO 980
830 WA=UD(LE)
840 WB=UP(LE)
850 AL=AS(LE)
860 X3=(L-AL)↑3
870 X2=(L-AL)↑2
880 RA=WA*X3*(L+AL)/(2*L↑3)
890 RA=RA+(WB-WA)*X3*(3*L+2*AL)/(20*L↑3)
900 RB=(WA+WB)*(L-AL)/2-RA
910 BA=-WA*X3*(L+3*AL)/(12*L↑2)
920 BA=BA-(WB-WA)*X3*(2*L+3*AL)/(60*L↑2)
930 BB=RA*L+BA-WA*X2/2-(WB-WA)*X2/6
940 Q(I1+1)=Q(I1+1)+RA
950 Q(J1+1)=Q(J1+1)+RB
960 Q(I1+2)=Q(I1+2)+BA
970 Q(J1+2)=Q(J1+2)-BB
980 FOR II=1 TO 4
990 FOR JJ=II TO 4
1000 MG=I1+II
1010 TR =I1+JJ-MG+1
1040 A(MG,TR)=A(MG,TR)+ST(II,JJ)
1080 NEXT JJ
1090 NEXT II
1100 NEXT LE
1110 IF NC=0 THEN 1200
1120 PRINT "POSITIONS AND VALUES OF CONCENTRATED LOADS"
1130 FOR II=1 TO NC
1140 PRINT "POSITION OF LOAD";II
1150 INPUT NP
1160 PRINT "VALUE OF CONCENTRATED LOAD";II
1170 INPUT QC
1180 Q(NP)=Q(NP)+QC
1190 NEXT II
1200 FOR II=1 TO NF
1210 N9=NS(II)
1220 A(N9,1)=A(N9,1)*1E12+1E12
1230 Q(N9)=0
1240 NEXT II
1242 FOR II=1 TO NN
1244 C(II)=Q(II)
1246 NEXTII
1255 N=NN
1260 GOSUB 6000
1270 FOR II=1 TO NN
1280 U(II)=C(II)
1320 NEXT II
1330 PRINT:PRINT:PRINT"THE NODAL DISPLACEMENTS V AND THETA ARE AS FOLLOWS":PRIN
T
1340 FOR II=1 TO ND
1350 PRINT U(2*II-1);U(2*II)
1360 NEXT II
1365 PRINT:PRINT:PRINT"THE NODAL BENDING MOMENTS ARE GIVEN BELOW";
1366 PRINT" - TO OBTAIN SUCCESSIVE BENDING MOMENT VALUES, PRESS ANY KEY":PRINT:
1370 FOR LE=1 TO LS
1380 I=LE
1390 J=I+1
1400 L=XL(LE)
1410 IF NL=0 THEN SW=UD(LE)
1420 JJ=2*LE-2
1430 FOR II=1 TO 4
1440 DI(II)=U(JJ+II)
1450 NEXT II
1460 XI=-1
1470 FOR IJ=1 TO 2
```

```
1480 XI=XI+1
1490 B(1)=-6+12*XI
1500 B(2)=L*(4-6*XI)
1510 B(3)=6-12*XI
1520 B(4)=L*(2-6*XI)
1530 CN =-1/L↑2*E*SA(LE)
1540 FOR KK=1 TO 4
1550 B(KK)=B(KK)*CN
1560 NEXT KK
1570 TH=0
1580 FOR KK=1 TO 4
1590 TH=TH+B(KK)*DI(KK)
1600 NEXT KK
1610 IF NL=1 THEN 1640
1620 TH=TH-SW*L↑2/12
1630 GOTO 1770
1640 WA=UD(LE)
1650 WB=UP(LE)
1660 AL=AS(LE)
1670 X3=(L-AL)↑3
1680 X2=(L-AL)↑2
1690 RA=WA*X3*(L+AL)/(2*L↑3)
1700 RA=RA+(WB-WA)*X3*(3*L+2*AL)/(20*L↑3)
1710 RB=(WA+WB)*(L-AL)/2-RA
1720 BA=-WA*X3*(L+3*AL)/(12*L↑2)
1730 BA=BA-(WB-WA)*X3*(2*L+3*AL)/(60*L↑2)
1740 BB=RA*L+BA-WA*X2/2-(WB-WA)*X2/6
1750 IF IJ=1 THEN TH=TH+BA
1760 IF IJ=2 THEN TH=TH+BB
1770 IF IJ=1 THEN II=I
1780 IF IJ=2 THEN II=J
1790 PRINT "NODE=";II;"MOMENT=";TH
1795 GETA$:IFA$=""THEN1795
1800 NEXT IJ
1810 NEXT LE
1830 END
6000 FORII=1TON
6010 IK=II
6020 FORJJ=2TONW
6030 IK=IK+1
6040 CN=A(II,JJ)/A(II,1)
6050 JK=0
6060 FORKK=JJTONW
6070 JK=JK+1
6080 A(IK,JK)=A(IK,JK)-CN*A(II,KK)
6090 NEXTKK
6100 A(II,JJ)=CN
6110 C(IK)=C(IK)-CN*C(II)
6120 NEXTJJ
6130 C(II)=C(II)/A(II,1)
6140 NEXTII
6150 FORIZ=2TON
6160 II=N-IZ+1
6170 FORKK=2TONW
6180 JJ=II+KK-1
6190 C(II)=C(II)-A(II,KK)*C(JJ)
6200 NEXTKK,IZ
6210 RETURN
READY.
```

APPENDIX 3

```
100 PRINT "STATIC ANALYSIS OF RIGID JOINTED PLANE FRAMES"
110 PRINT "FEED IN THE NO OF NODAL POINTS"
120 INPUT NJ
130 N3=NJ*3
140 PRINT"NO OF ELEMENTS"
150 INPUT MS
160 PRINT "NO OF SUPPRESSED DISPLTS"
170 INPUT NF
180 PRINT "NO OF CONCENTRATED LOADS"
190 INPUT NC
200 DIM U(N3),K(6,6),NS(NF),I9(MS),J9(MS)
210 DIM DC(6,6),SP(6),SQ(6),SG(6),SL(6),X(NJ),Y(NJ)
220 DIM S9(MS),A9(MS),U9(MS),UP(MS),AS(MS),BX(2,6),S2(2)
230 PRINT "NODAL COORDINATES"
240 FOR I=1 TO NJ
250 PRINT "X COORD FOR NODE";I
260 INPUT X(I)
270 PRINT "Y COORD FOR NODE";I
280 INPUT Y(I)
290 NEXT I
300 PRINT "POSNS OF SUPPRESSED DISPLTS"
310 FOR I=1 TO NF
320 PRINT "POSITION";I
330 INPUT NS(I):NEXT I
340 PRINT "ELASTIC MODULUS"
350 INPUT E
360 PRINT "IF HYDROSTATIC LOAD FEED 1 ELSE FEED 0"
370 INPUT HY
380 MX=0
390 IF HY=0 THEN 560
400 FOR I=1 TO MS:PRINT"FEED I NODE FOR ELEM";I:INPUT I9(I)
410 PRINT"FEED J NODE FOR ELEM";I:INPUT J9(I)
420 I9=I9(I):J9=J9(I)
430 IF ABS(J9-I9)>MXTHEN MX=ABS(J9-I9)
440 PRINT "2ND MOM OF AREA FOR ELEM";I
450 INPUT S9(I)
460 PRINT "CROSS SECTNL AREA FOR ELEM";I
470 INPUT A9(I)
480 PRINT "VALUE OF HYDRO LOAD (ON LEFT) FOR ELEM";I
490 INPUT U9(I)
500 PRINT "VALUE OF HYDRO LOAD (ON RIGHT) FOR ELEM";I
510 INPUT UP(I)
520 PRINT "FEED IN THE DISTANCE (A) FOR ELEM";I
530 INPUT AS(I)
540 NEXT I
550 GOTO 660
560 FOR I=1 TO MS:PRINT"FEED I NODE FOR ELEM";I:INPUT I9(I)
570 PRINT"FEED J NODE FOR ELEM";I:INPUT J9(I)
580 I9=I9(I):J9=J9(I)
590 IF ABS(J9-I9)>MXTHEN MX=ABS(J9-I9)
600 PRINT "2ND MOM OF AREA FOR ELEM";I
610 INPUT S9(I)
620 PRINT "CROSS SECTNL AREA FOR ELEM";I
630 INPUT A9(I)
640 PRINT "UDL FOR ELEMENT";I
650 INPUT U9(I):NEXT I
660 NW=(MX+1)*3:NT=N3+NW
670 DIMA(NT,NW),Q(NT),C(NT)
680 FOR I=1 TO NT:FOR J=1 TO NW:A(I,J)=0
690 NEXTJ:Q(I)=0:NEXTI
700 FOR ME =1 TO MS
705 PRINT:PRINT"ELEMENT";ME;"UNDER COMPUTATION":PRINT
710 I=I9(ME)
720 J=J9(ME)
730 SA=S9(ME)
740 CA=A9(ME)
750 IF HY=0 THEN UD=U9(ME)
760 L=SQR((X(J)-X(I))↑2+(Y(J)-Y(I))↑2)
770 C=(X(J)-X(I))/L
780 S=(Y(J)-Y(I))/L
790 I3=3*I
800 I2=I3-1
```

```
810 I1=I3-2
820 J3=3*J
830 J2=J3-1
840 J1=J3-2
850 C1=12*E*SA*S↑2/L↑3+C↑2*CA*E/L
860 C2=12*E*SA*C*S/L↑3-C*S*CA*E/L
870 C3=12*E*SA*C↑2/L↑3+S↑2*CA*E/L
880 C4=6*E*SA*S/L↑2
890 C5=6*E*SA*C/L↑2
900 C6=4*E*SA/L
910 K(1,1)=C1
920 K(2,1)=-C2
930 K(1,2)=-C2
940 K(3,1)=C4
950 K(1,3)=C4
960 K(4,1)=-C1
970 K(1,4)=-C1
980 K(5,1)=C2
990 K(1,5)=C2
1000 K(6,1)=C4
1010 K(1,6)=C4
1020 K(2,2)=C3
1030 K(3,2)=-C5
1040 K(2,3)=-C5
1050 K(4,2)=C2
1060 K(2,4)=C2
1070 K(5,2)=-C3
1080 K(2,5)=-C3
1090 K(6,2)=-C5
1100 K(2,6)=-C5
1110 K(3,3)=C6
1120 K(4,3)=-C4
1130 K(3,4)=-C4
1140 K(5,3)=C5
1150 K(3,5)=C5
1160 K(6,3)=0.5*C6
1170 K(3,6)=0.5*C6
1180 K(4,4)=C1
1190 K(5,4)=-C2
1200 K(4,5)=-C2
1210 K(6,4)=-C4
1220 K(4,6)=-C4
1230 K(5,5)=C3
1240 K(6,5)=C5
1250 K(5,6)=C5
1260 K(6,6)=C6
1270 SP(1)=0
1280 SP(4)=0
1290 IF HY=1 THEN 1350
1300 SP(2)=UD*L/2
1310 SP(5)=UD*L/2
1320 SP(3)=-UD*L↑2/12
1330 SP(6)=UD*L↑2/12
1340 GOTO 1480
1350 WA=U9(ME)
1360 WB=UP(ME)
1370 AL=AS(ME)
1380 X3=(L-AL)↑3
1390 X2=(L-AL)↑2
1400 SP(2)=WA*X3*(L+AL)/(2*L↑3)
1410 SP(2)=SP(2)+(WB-WA)*X3*(3*L+2*AL)/(20*L↑3)
1420 SP(5)=(WA+WB)*(L-AL)/2-SP(2)
1430 SP(3)=-WA*X3*(L+3*AL)/(12*L↑2)
1440 SP(3)=SP(3)-(WB-WA)*X3*(2*L+3*AL)/(60*L↑2)
1450 SP(6)=SP(2)*L+SP(3)-WA*X2/2
1460 SP(6)=SP(6)-(WB-WA)*X2/6
1470 SP(6)=-SP(6)
1480 FOR II=1 TO 6
1490 FOR JJ=1 TO 6
1500 DC(II,JJ)=0
1510 NEXT JJ
1520 NEXT II
1530 DC(1,1)=C
1540 DC(2,2)=C
1550 DC(4,4)=C
```

```
1560 DC(5,5)=C
1570 DC(2,1)=S
1580 DC(5,4)=S
1590 DC(1,2)=-S
1600 DC(4,5)=-S
1610 DC(3,3)=1
1620 DC(6,6)=1
1630 FOR II=1 TO 6
1640 SQ(II)=0
1650 FOR JJ=1 TO 6
1660 SQ(II)=SQ(II)+DC(II,JJ)*SP(JJ)
1670 NEXT JJ
1680 NEXT II
1690 Q(I1)=Q(I1)+SQ(1)
1700 Q(I2)=Q(I2)+SQ(2)
1710 Q(I3)=Q(I3)+SQ(3)
1720 Q(J1)=Q(J1)+SQ(4)
1730 Q(J2)=Q(J2)+SQ(5)
1740 Q(J3)=Q(J3)+SQ(6)
1750 I1=3*I-3
1760 J1=3*J-3
1770 FORJJ=1TO2
1772 IFJJ=1THENNR=I1
1774 IFJJ=2THENNR=J1
1780 FORJ9=1TO3
1785 NR=NR+1:II=(JJ-1)*3+J9
1790 FORKK=1TO2
1792 IFKK=1THENN9=I1
1794 IFKK=2THENN9=J1
1800 FORK=1TO3
1805 LL=(KK-1)*3+K:NK=N9+K+1-NR
1810 IF NK<=0 THEN 1830
1820 A(NR,NK)=A(NR,NK)+K(II,LL)
1830 NEXTK
1835 NEXTKK,J9,JJ
1840 NEXT ME
1850 IF NC=0 THEN 1930
1855 PRINT:PRINT"FEED IN THE (POSITIONS) & VALUES OF THE CONCENTRATED (LOADS)":
PRINT
1860 FOR I=1 TO NC
1870 PRINT "(POSITION) OF CONCENTRATED LOAD"
1880 INPUT PO
1890 PRINT "VALUE OF CONCENTRATED LOAD"
1900 INPUT QC
1910 Q(PO)=Q(PO)+QC
1920 NEXT I
1930 FOR I=1 TO NF
1940 A(NS(I),1)=A(NS(I),1)*1E12+1E12
1945 Q(NS(I))=0
1950 NEXT I
1960 FORII=1TON3
1970 C(II)=Q(II)
1980 NEXTII
1990 N=N3
1995 PRINT:PRINT"THE SIMULTANEOUS EQUATIONS ARE NOW BEING SOLVED":PRINT
2000 GOSUB 2940
2010 FOR I=1 TO N3
2020 U(I)=C(I)
2030 NEXTI
2040 PRINT "THE NODAL DISPLACEMENTS ARE":PRINT
2050 FOR I=1 TO N3
2060 PRINT U(I);
2070 NEXT I
2080 FOR ME=1 TO MS
2090 I=I9(ME)
2100 J=J9(ME)
2110 SA=S9(ME)
2120 CA=A9(ME)
2130 L=SQR((X(J)-X(I))↑2+(Y(J)-Y(I))↑2)
2140 C=(X(J)-X(I))/L
2150 S=(Y(J)-Y(I))/L
2160 IF HY=1 THEN 2190
2170 UD=U9(ME)
2180 GOTO 2290
2190 WA=U9(ME)
```

```
2200 WB=UP(ME)
2210 AL=AS(ME)
2220 X3=(L-AL)↑3
2230 X2=(L-AL)↑2
2240 SP(2)=WA*X3*(L+AL)/(2*L↑3)
2250 SP(2)=SP(2)+(WB-WA)*X3*(3*L+2*AL)/(20*L↑3)
2260 SP(3)=-WA*X3*(L+3*AL)/(12*L↑2)
2270 SP(3)=SP(3)-(WB-WA)*X3*(2*L+3*AL)/(60*L↑2)
2280 SP(6)=SP(2)*L+SP(3)-WA*X2/2-(WB-WA)*X2/6
2290 I1=3*I-3
2300 J1=3*J-3
2310 FOR II=1 TO 3
2320 MM=I1+II
2330 MN=J1+II
2340 SG(II)=U(MM)
2350 SG(II+3)=U(MN)
2360 NEXT II
2370 FOR II=1 TO 6
2380 FOR JJ=1 TO 6
2390 DC(II,JJ)=0
2400 NEXT JJ
2410 NEXT II
2420 DC(1,1)=C
2430 DC(2,2)=C
2440 DC(4,4)=C
2450 DC(5,5)=C
2460 DC(1,2)=S
2470 DC(4,5)=S
2480 DC(2,1)=-S
2490 DC(5,4)=-S
2500 DC(3,3)=1
2510 DC(6,6)=1
2520 FOR II=1 TO 6
2530 SL(II)=0
2540 FOR JJ=1 TO 6
2550 SL(II)=SL(II)+DC(II,JJ)*SG(JJ)
2560 NEXT JJ
2570 NEXT II
2580 PRINT
2590 FOR II=1 TO 2
2600 IF II=1 THEN XI=0
2610 IF II=1 THEN ND=I
2620 IF II=2 THEN XI=1
2630 IF II=2 THEN ND=J
2640 FOR JJ=1 TO 2
2650 FOR KK=1 TO 6
2660 BX(JJ,KK)=0
2670 NEXT KK
2680 NEXT JJ
2690 BX(1,1)=-1/L
2700 BX(1,4)=1/L
2710 BX(2,2)=(6-12*XI)/L↑2
2720 BX(2,3)=(-4+6*XI)/L
2730 BX(2,5)=-(6-12*XI)/L↑2
2740 BX(2,6)=(-2+6*XI)/L
2750 FOR JJ=1 TO 2
2760 S2(JJ)=0
2770 FOR KK=1 TO 6
2780 S2(JJ)=S2(JJ)+BX(JJ,KK)*SL(KK)
2790 NEXT KK
2800 NEXT JJ
2810 S2(1)=E*CA*S2(1)
2820 S2(2)=E*SA*S2(2)
2830 IF HY=1 THEN 2860
2840 S2(2)=S2(2)-UD*L↑2/12
2850 GOTO 2880
2860 IF II=1 THEN S2(2)=S2(2)+SP(3)
2870 IF II=2 THEN S2(2)=S2(2)+SP(6)
2880 PRINT "FORCE IN ELEMENT";ME;"AT NODE";ND;S2(1)
2890 PRINT "BM IN ELEMENT";ME;"AT NODE";ND;S2(2)
2895 PRINT:PRINT"TO CONTINUE PRESS ANY KEY":PRINT
2900 GET A$:IF A$="" GOTO 2900
2910 NEXT II
2920 NEXT ME
2930 END
```

```
 2940 FORII=1 TO N
 2950 IK=II
 2960 FORJJ=2 TO NW
 2970 IK=IK+1
 2980 CN=A(II,JJ)/A(II,1)
 2990 JK=0
 3000 FOR KK=JJ TO NW
 3010 JK=JK+1
 3020 A(IK,JK)=A(IK,JK)-CN*A(II,KK)
 3030 NEXT KK
 3040 A(II,JJ)=CN
 3050 C(IK)=C(IK)-CN*C(II)
 3060 NEXT JJ
 3070 C(II)=C(II)/A(II,1)
 3080 NEXT II
 3090 FORIZ=2TON
 3100 II=N-IZ+1
 3110 FORKK=2TONW
 3120 JJ=II+KK-1
 3130 C(II)=C(II)-A(II,KK)*C(JJ)
 3140 NEXTKK,IZ
 3150 RETURN
READY.
```

APPENDIX 4

```
180 PRINT "FORCES IN PIN JOINTED SPACE TRUSSES"
190 PRINT "NO OF PIN JOINTS"
200 INPUT NJ
210 PRINT "NO OF MEMBERS"
220 INPUT MS
230 PRINT "NO OF SUPPRESSED DISPLTS"
240 INPUT NF
250 N3=NJ*3
260 N=N3-NF
270 DIM U(N3),AA(MS),EA(MS),IA(MS),JA(MS)
280 DIM ST(6,6),X(NJ),Y(NJ),Z(NJ),NS(NF),SG(3),SH(3),DC(1,3)
300 PRINT "NODAL COORDINATES"
310 FOR I=1 TO NJ
320 PRINT "X COORD FOR NODE";I
330 INPUT X(I)
340 PRINT "Y COORD FOR NODE";I
350 INPUT Y(I)
360 PRINT "Z COORD FOR NODE";I
370 INPUT Z(I)
380 NEXT I
390 PRINT "POSNS OF SUPPRESSED DISPLTS"
400 FOR I=1 TO NF
410 PRINT "POSITION";I
420 INPUT NS(I)
430 NEXT I
435 MX=0
440 PRINT "MEMBER DETAILS"
450 FOR I=1 TO MS
460 PRINT "I NODE FOR MEMBER";I
470 INPUT IA(I)
480 PRINT "J NODE FOR MEMBER";I
490 INPUT JA(I)
500 PRINT "CROSS SECTNL AREA FOR MEM";I
510 INPUT AA(I)
520 PRINT "E FOR MEMBER";I
530 INPUT EA(I)
532 I9=IA(I):J9=JA(I)
534 IF ABS(J9-I9)>MX THEN MX=ABS(J9-I9)
540 NEXT I
600 NW=(MX+1)*3:NT=N3+NW
610 DIM A(NT,NW),Q(NT),C(NT)
620 FOR I=1 TO NT:FOR J=1 TO NW:A(I,J)=0
630 NEXT J:Q(I)=0:NEXT I
635 PRINT "VECTOR OF EXTERNAL LOADS"
640 FOR I=1 TO N3
645 PRINT "POSITION";I
650 INPUT Q(I)
655 NEXT I
660 FOR ME=1 TO MS
670 PRINT"MEMBER ";ME;" UNDER COMPUTATION"
680 I=IA(ME)
690 J=JA(ME)
700 A=AA(ME)
710 E=EA(ME)
720 L=SQR((X(J)-X(I))↑2+(Y(J)-Y(I))↑2+(Z(J)-Z(I))↑2)
730 XC=(X(J)-X(I))/L
740 YC=(Y(J)-Y(I))/L
750 ZC=(Z(J)-Z(I))/L
760 ST(1,1)=XC↑2
770 ST(1,2)=XC*YC
780 ST(2,1)=XC*YC
790 ST(2,2)=YC↑2
800 ST(1,3)=XC*ZC
810 ST(3,1)=XC*ZC
820 ST(2,3)=YC*ZC
830 ST(3,2)=YC*ZC
840 ST(3,3)=ZC↑2
850 ST(4,1)=-XC↑2
860 ST(4,2)=-XC*YC
870 ST(4,3)=-XC*ZC
880 ST(5,1)=-XC*YC
890 ST(5,2)=-YC↑2
```

```
 900 ST(5,3)=-YC*ZC
 910 ST(6,1)=-XC*ZC
 920 ST(6,2)=-YC*ZC
 930 ST(6,3)=-ZC^2
 940 FOR II=1 TO 3
 950 FOR JJ=1 TO 3
 960 ST(II+3,JJ+3)=ST(II,JJ)
 970 ST(II,JJ+3)=ST(JJ+3,II)
 980 NEXT JJ
 990 NEXT II
 1000 CN=A*E/L
 1010 FOR II=1 TO 6
 1020 FOR JJ=1 TO 6
 1030 ST(II,JJ)=ST(II,JJ)*CN
 1040 NEXT JJ
 1050 NEXT II
 1060 I1=3*I-3
 1070 J1=3*J-3
 1075 FOR JJ=1 TO 2
 1080 IF JJ=1 THEN NR=I1
 1085 IF JJ=2 THEN NR=J1
 1090 FOR J9=1 TO 3
 1095 NR=NR+1:II=(JJ-1)*3+J9
 1100 FOR KK=1 TO 2
 1105 IF KK=1 THEN N9=I1
 1110 IF KK=2 THEN N9=J1
 1115 FOR K=1 TO 3
 1120 LL=(KK-1)*3+K
 1125 NK=N9+K+1-NR
 1130 IF NK<=0 THEN 1140
 1135 A(NR,NK)=A(NR,NK)+ST(II,LL)
 1140 NEXT K
 1150 NEXT KK,J9,JJ
 1200 NEXT ME
 1210 FOR I=1 TO NF
 1220 A(NS(I),1)=A(NS(I),1)*1E12+1E12
 1225 Q(NS(I))=0
 1230 NEXT I
 1240 N=N3
 1250 FOR II=1 TO N3:C(II)=Q(II):NEXT II
 1255 PRINT:PRINT:PRINT"SOLUTION OF SIMULTANEOUS EQUATIONS BEING CARRIED OUT":PR
INT:PRINT
 1260 GOSUB 2940
 1270 FOR II=1 TO N3
 1290 U(II)=C(II)
 1310 NEXT II
 1320 PRINT
 1330 PRINT "NODAL DISPLACEMENTS"
 1340 PRINT
 1350 FOR II=1 TO N3
 1360 PRINT U(II);
 1370 NEXT II
 1380 PRINT
 1390 PRINT "FORCES IN THE MEMBERS":PRINT
 1400 FOR ME=1 TO MS
 1410 I=IA(ME)
 1420 J=JA(ME)
 1430 A=AA(ME)
 1440 E=EA(ME)
 1450 L=SQR((X(J)-X(I))^2+(Y(J)-Y(I))^2+(Z(J)-Z(I))^2)
 1460 XC=(X(J)-X(I))/L
 1470 YC=(Y(J)-Y(I))/L
 1480 ZC=(Z(J)-Z(I))/L
 1490 DC(1,1)=XC
 1500 DC(1,2)=YC
 1510 DC(1,3)=ZC
 1520 I1=3*I-3
 1530 J1=3*J-3
 1540 FOR I3=1 TO 3
 1550 J3=I1+I3
 1560 J2=J1+I3
 1570 SG(I3)=U(J3)
 1580 SH(I3)=U(J2)
 1590 NEXT I3
 1600 A1=0
```

```
1610 A2=0
1620 FOR II=1 TO 3
1630 A1=A1+DC(1,II)*SG(II)
1640 A2=A2+DC(1,II)*SH(II)
1650 NEXT II
1660 FC=A*E*(A2-A1)/L
1670 PRINT "FORCE IN MEM";ME;"=";FC
1675 PRINT:PRINT"TO CONTINUE PRESS ANY KEY":PRINT
1680 GET A$:IF A$="" THEN GOTO 1680
1690 NEXT ME
1700 END
2940 FOR II=1 TO N
2950 IK=II
2960 FOR JJ=2 TO NW
2970 IK=IK+1
2980 CN=A(II,JJ)/A(II,1)
2990 JK=0
3000 FOR KK=JJ TO NW
3010 JK=JK+1
3020 A(IK,JK)=A(IK,JK)-CN*A(II,KK)
3030 NEXT KK
3040 A(II,JJ)=CN
3050 C(IK)=C(IK)-CN*C(II)
3060 NEXT JJ
3070 C(II)=C(II)/A(II,1)
3080 NEXT II
3090 FORIZ=2TON
3100 II=N-IZ+1
3110 FORKK=2TONW
3120 JJ=II+KK-1
3130 C(II)=C(II)-A(II,KK)*C(JJ)
3140 NEXTKK,IZ
3150 RETURN
READY.
```

APPENDIX 5

```
10 PRINT:PRINT:PRINT"FREE VIBRATION OF PLANE PIN JOINTED TRUSSES":PRINT:PRINT
20 PRINT "NO OF NODAL POINTS"
30 INPUT NN
40 PRINT "NO OF MEMBERS"
50 INPUT LS
60 PRINT "NO OF SUPPRESSED DISPLTS"
70 INPUT NF
80 N2=NN*2
90 NP=N2+1
100 N=N2-NF
110 PRINT "NO OF FREQUENCIES-THIS MUST BE <=";N
120 INPUT M1
130 DIM A(N2,N2),XX(N2,NP),AM(M1),VC(N,M1),X(40),Z(N),IG(N)
140 DIM ST(4,4),SM(4,4),NS(NF),Y(NN),VE(N)
150 GOTO 6020
200 PRINT:PRINT"THE EIGENVALUES ARE BEING DETERMINED":PRINT
210 GOSUB 500
220 FOR I=1 TO M1
240 PRINT "EIGENVALUE=";AM(I)
250 PRINT "FREQUENCY=";SQR(1/AM(I))/(2*π)
260 PRINT "EIGENVECTOR IS"
270 PRINT "    "
280 FOR J=1 TO N
300 PRINT VC(J,I);" ";
320 NEXT J
325 PRINT:PRINT:PRINT"TO CONTINUE, PRESS ANY KEY"
330 GET A$:IF A$="" THEN 330
340 NEXT I
360 END
500 MN=N
510 NN=N
520 GOSUB 1500
530 M=1
540 FOR I=1 TO NN
550 VC(I,M)=X(I)
560 XX(I,M)=VC(I,M)
570 NEXT I
580 AM(M)=XM
590 IF M1<2 THEN 220
600 FOR M=2 TO M1
610 FOR I=1 TO NN
620 K4=ABS(XX(I,M-1)-1)
630 IF K4<0.00001 THEN IR=I
650 NEXT I
660 IG(M-1)=IR
670 FOR I=1 TO NN
680 XX(MN-I+1,MN-M+3)=A(IR,I)
690 NEXT I
700 FOR I=1 TO NN
710 FOR J=1 TO NN
720 Z1=MN-J+1
730 Z2=MN-M+3
740 A(I,J)=A(I,J)-XX(I,M-1)*XX(Z1,Z2)
750 NEXT J
760 NEXT I
770 FOR I=1 TO NN
780 IF I=IR THEN 910
790 IF I>IR THEN K1=I-1
810 IF I<=IR THEN K1=I
830 FOR J=1 TO NN
840 IF J=IR THEN 900
850 IF J>IR THEN K2=J-1
870 IF J<=IR THEN K2=J
890 A(K1,K2)=A(I,J)
900 NEXT J
910 NEXT I
920 NN=NN-1
930 M3=NN
940 IF M<>MN THEN 980
950 XM=A(1,1)
960 X(1)=1
970 GOTO 990
```

```
980 GOSUB 1500
990 FOR I=1 TO NN
1000 XX(I,M)=X(I)
1010 NEXT I
1020 AM(M)=XM
1030 M4=M-1
1040 M5=1000-M4
1050 FOR M8=M5 TO 999
1060 M6=M3+1
1070 M2=1000-M8
1080 M7=IG(M2)+1
1090 IF M6<M7 THEN 1150
1100 N9=1000-M7
1110 N8=1000-M6
1120 FOR I3=N8 TO N9
1125 I=1000-I3
1130 X(I)=X(I-1)
1140 NEXT I3
1150 J=IG(M2)
1160 X(J)=0
1170 SM=0
1180 FOR I=1 TO M6
1190 Z3=MN-I+1
1200 Z4=MN-M2+2
1210 SM=SM+XX(Z3,Z4)*X(I)
1220 NEXT I
1230 XK=(AM(M2)-XM)/SM
1240 FOR I=1 TO M6
1250 X(I)=XX(I,M2)-XK*X(I)
1260 NEXT I
1270 SM=0
1280 FOR I=1 TO M6
1290 IF ABS(SM)<ABS(X(I)) THEN SM=X(I)
1310 NEXT I
1320 FOR I=1 TO M6
1330 X(I)=X(I)/SM
1340 NEXT I
1350 M3=M3+1
1360 IF M2<>1 THEN 1400
1370 FOR I=1 TO M3
1380 VC(I,M)=X(I)
1390 NEXT I
1400 NEXT M8
1405 NEXT M
1410 RETURN
1420 END
1500 Y1=100000
1510 FOR I=1 TO NN
1520 X(I)=1
1530 NEXT I
1540 XM=-100000
1550 FOR I=1 TO NN
1560 SG=0
1570 FOR J=1 TO NN
1580 SG=SG+A(I,J)*X(J)
1590 NEXT J
1600 Z(I)=SG
1610 NEXT I
1620 XM=0
1630 FOR I=1 TO NN
1640 IF ABS(XM)<ABS(Z(I)) THEN XM=Z(I)
1650 NEXT I
1660 FOR I=1 TO NN
1670 X(I)=Z(I)/XM
1680 NEXT I
1690 IF ABS((Y1-XM)/XM)>D THEN 1710
1700 GOTO 1730
1710 Y1=XM
1720 GOTO 1550
1730 X3=0
1740 FOR I=1 TO NN
1750 IF ABS(X3)<ABS(X(I)) THEN X3=X(I)
1760 NEXT I
1770 FOR I=1 TO NN
1780 X(I)=X(I)/X3
```

```
1790 NEXT I
1800 RETURN
1810 END
3200 N9=N-1
3210 FORX=1TON
3220 DI=A(X,1)
3230 IFDI=0THENPRINT"MATRIX IS SINGULAR"
3240 FORY=1TON9
3250 Y9=Y+1
3260 A(X,Y)=A(X,Y9)/DI
3270 NEXTY
3280 A(X,N)=1/DI
3290 FORZ=1TON
3300 IFZ=XTHEN3370
3310 O=A(Z,1)
3320 FORY=1TON9
3330 Y9=Y+1
3340 A(Z,Y)=A(Z,Y9)-A(X,Y)*O
3350 NEXTY
3360 A(Z,N)=-A(X,N)*O
3370 NEXTZ
3380 NEXTX
3390 RETURN
6020 FOR I=1 TO N2
6030 FOR J=1 TO N2
6040 A(I,J)=0.0
6050 NEXT J
6060 FOR J=1 TO N2+1
6070 XX(I,J)=0.0
6080 NEXT J
6090 NEXT I
6100 PRINT "DISPLACEMENT POSNS OF SUPPRESSED DISPLTS - IN ASCENDING ORDER"
6120 FOR I=1 TO NF
6130 PRINT "POSITION";I
6140 INPUT NS(I)
6150 NEXT I
6160 PRINT "FEED IN NODAL COORDINATES"
6170 FOR I=1 TO NN
6180 PRINT "X COORDINATE OF NODE";I
6190 INPUT X(I)
6200 PRINT "Y COORDINATE OF NODE";I
6210 INPUT Y(I)
6220 NEXT I
6230 PRINT "FEED IN MEMBER DETAILS"
6240 FOR LE =1 TO LS
6245 PRINT"MEMBER NO";LE
6250 PRINT "I NODE"
6260 INPUT I
6270 PRINT "J NODE"
6280 INPUT J
6290 PRINT "CROSS SECTIONAL AREA"
6300 INPUT A
6310 PRINT "ELASTIC MODULUS"
6320 INPUT E
6330 PRINT "DENSITY"
6340 INPUT RH
6345 PRINT:PRINT"MEMBER NO";LE;"UNDER COMPUTATION":PRINT
6350 L=SQR((X(J)-X(I))↑2+(Y(J)-Y(I))↑2)
6360 C=(X(J)-X(I))/L
6370 S=(Y(J)-Y(I))/L
6380 ST(1,1)=C↑2
6390 ST(1,2)=C*S
6400 ST(1,3)=-C↑2
6410 ST(1,4)=-C*S
6420 ST(2,1)=C*S
6430 ST(2,2)=S↑2
6440 ST(2,3)=-C*S
6450 ST(2,4)=-S↑2
6460 ST(3,1)=-C↑2
6470 ST(3,2)=-C*S
6480 ST(3,3)=C↑2
6490 ST(3,4)=C*S
6500 ST(4,1)=-C*S
6510 ST(4,2)=-S↑2
6520 ST(4,3)=C*S
```

```
6530 ST(4,4)=S↑2
6540 FOR II=1 TO 4
6550 FOR JJ=1 TO 4
6560 SM(II,JJ)=0.0
6570 NEXT JJ
6580 NEXT II
6590 SM(1,1)=2
6600 SM(2,2)=2
6610 SM(3,3)=2
6620 SM(4,4)=2
6630 SM(1,3)=1
6640 SM(3,1)=1
6650 SM(2,4)=1
6660 SM(4,2)=1
6670 FOR II=1 TO 4
6680 FOR JJ=1 TO 4
6690 CN=A*E/L
6700 ST(II,JJ)=ST(II,JJ)*CN
6710 CN=RH*A*L/6
6720 SM(II,JJ)=SM(II,JJ)*CN
6730 NEXT JJ
6740 NEXT II
6750 I1=2*I-2
6760 J1=2*J-2
6770 FOR II=1 TO 2
6780 FOR JJ=1 TO 2
6790 MM=I1+II
6800 MN=I1+JJ
6810 NM=J1+II
6820 N9=J1+JJ
6830 A(MM,MN)=A(MM,MN)+ST(II,JJ)
6840 XX(MM,MN)=XX(MM,MN)+SM(II,JJ)
6850 A(NM,MN)=A(NM,MN)+ST(II+2,JJ)
6860 XX(NM,MN)=XX(NM,MN)+SM(II+2,JJ)
6870 A(MM,N9)=A(MM,N9)+ST(II,JJ+2)
6880 XX(MM,N9)=XX(MM,N9)+SM(II,JJ+2)
6890 A(NM,N9)=A(NM,N9)+ST(II+2,JJ+2)
6900 XX(NM,N9)=XX(NM,N9)+SM(II+2,JJ+2)
6910 NEXT JJ
6920 NEXT II
6930 NEXT LE
6940 MM=0
6950 FOR I=1 TO NF
6960 N7=NS(I)
6970 MM=MM+1
6980 N7=N7-MM+1
6990 N8=N2-MM
7000 IF N8<N7 THEN 7070
7010 FOR II=N7 TO N8
7020 FOR JJ=N7 TO N8
7030 A(II,JJ)=A(II+1,JJ+1)
7040 XX(II,JJ)=XX(II+1,JJ+1)
7050 NEXT JJ
7060 NEXT II
7070 IF N7=1 THEN 7160
7080 FOR II=1 TO N7-1
7090 FOR JJ=N7 TO N8
7100 A(II,JJ)=A(II,JJ+1)
7110 XX(II,JJ)=XX(II,JJ+1)
7120 A(JJ,II)=A(JJ+1,II)
7130 XX(JJ,II)=XX(JJ+1,II)
7140 NEXT JJ
7150 NEXT II
7160 NEXT I
7190 PRINT "NO OF CONCENTRATED MASSES"
7200 INPUT NC
7210 IF NC=0 THEN 7345
7220 FOR II=1 TO NC
7230 PRINT "NODAL POSN OF MASS"
7240 INPUT PO
7250 PRINT "VALUE OF MASS"
7260 INPUT MC
7270 I1=2*PO-1
7275 I9=2*PO-1
7280 FOR JJ=1 TO NF
```

```
 7290 IF NS(JJ)<I9 THEN I1=I1-1
 7300 NEXT JJ
 7310 J1=I1+1
 7320 XX(I1,I1)=XX(I1,I1)+MC
 7330 XX(J1,J1)=XX(J1,J1)+MC
 7340 NEXT II
 7345 PRINT:PRINT"THE STIFFNESS MATRIX IS BEING INVERTED":PRINT
 7348 GOSUB 3200
 7350 FOR I=1 TO N
 7360 FOR J=1 TO N
 7370 VE(J)=0
 7380 FOR K=1 TO N
 7390 VE(J)=VE(J)+A(I,K)*XX(K,J)
 7400 NEXT K
 7410 NEXT J
 7420 FOR J=1 TO N
 7430 A(I,J)=VE(J)
 7440 NEXT J
 7450 NEXT I
 7460 D=0.001
 7470 GOTO 200
 7480 END
READY.
```

APPENDIX 6

```
10 PRINT:PRINT:PRINT
15 PRINT"FREE VIBRATION OF CONTINUOUS BEAMS OF VARYING CROSS SECTION":PRINT
20 PRINT "NO OF ELEMENTS"
30 INPUT LS
40 PRINT "NO OF SUPPRESSED DISPLTS"
50 INPUT NF
60 NN=2*LS+2
70 N=NN-NF
80 NP=NN+1
90 PRINT "NO OF FREQUENCIES REQD-THIS MUST BE <=";N
100 INPUT M1
110 DIM A(NN,NN),XX(NN,NP),VC(N,M1),AM(M1),X(N),Z(N),IG(N),NS(NF)
120 DIM ST(4,4),SM(4,4),VE(N)
150 GOTO6020
200 PRINT:PRINT"THE EIGENVALUES ARE BEING DETERMINED":PRINT
210 GOSUB 500
220 FOR I=1 TO M1
240 PRINT "EIGENVALUE=";AM(I)
250 PRINT "FREQUENCY=";SQR(1/AM(I))/(2*π)
260 PRINT "EIGENVECTOR IS"
270 PRINT "    "
280 FOR J=1 TO N
300 PRINT VC(J,I);" ";
320 NEXT J
325 PRINT:PRINT:PRINT"TO CONTINUE, PRESS ANY KEY":PRINT
330 GET A$:IF A$="" THEN 330
340 NEXT I
360 END
500 MN=N
510 NN=N
520 GOSUB 1500
530 M=1
540 FOR I=1 TO NN
550 VC(I,M)=X(I)
560 XX(I,M)=VC(I,M)
570 NEXT I
580 AM(M)=XM
590 IF M1<2 THEN 220
600 FOR M=2 TO M1
610 FOR I=1 TO NN
620 K4=ABS(XX(I,M-1)-1)
630 IF K4<0.00001 THEN IR=I
650 NEXT I
660 IG(M-1)=IR
670 FOR I=1 TO NN
680 XX(MN-I+1,MN-M+3)=A(IR,I)
690 NEXT I
700 FOR I=1 TO NN
710 FOR J=1 TO NN
720 Z1=MN-J+1
730 Z2=MN-M+3
740 A(I,J)=A(I,J)-XX(I,M-1)*XX(Z1,Z2)
750 NEXT J
760 NEXT I
770 FOR I=1 TO NN
780 IF I=IR THEN 910
790 IF I>IR THEN K1=I-1
810 IF I<=IR THEN K1=I
830 FOR J=1 TO NN
840 IF J=IR THEN 900
850 IF J>IR THEN K2=J-1
870 IF J<=IR THEN K2=J
890 A(K1,K2)=A(I,J)
900 NEXT J
910 NEXT I
920 NN=NN-1
930 M3=NN
940 IF M<>MN THEN 980
950 XM=A(1,1)
960 X(1)=1
970 GOTO 990
980 GOSUB 1500
```

```
990 FOR I=1 TO NN
1000 XX(I,M)=X(I)
1010 NEXT I
1020 AM(M)=XM
1030 M4=M-1
1040 M5=1000-M4
1050 FOR M8=M5 TO 999
1060 M6=M3+1
1070 M2=1000-M8
1080 M7=IG(M2)+1
1090 IF M6<M7 THEN 1150
1100 N9=1000-M7
1110 N8=1000-M6
1120 FOR I3=N8 TO N9
1125 I=1000-I3
1130 X(I)=X(I-1)
1140 NEXT I3
1150 J=IG(M2)
1160 X(J)=0
1170 SM=0
1180 FOR I=1 TO M6
1190 Z3=MN-I+1
1200 Z4=MN-M2+2
1210 SM=SM+XX(Z3,Z4)*X(I)
1220 NEXT I
1230 XK=(AM(M2)-XM)/SM
1240 FOR I=1 TO M6
1250 X(I)=XX(I,M2)-XK*X(I)
1260 NEXT I
1270 SM=0
1280 FOR I=1 TO M6
1290 IF ABS(SM)<ABS(X(I)) THEN SM=X(I)
1310 NEXT I
1320 FOR I=1 TO M6
1330 X(I)=X(I)/SM
1340 NEXT I
1350 M3=M3+1
1360 IF M2<>1 THEN 1400
1370 FOR I=1 TO M3
1380 VC(I,M)=X(I)
1390 NEXT I
1400 NEXT M8
1405 NEXT M
1410 RETURN
1420 END
1500 Y1=100000
1510 FOR I=1 TO NN
1520 X(I)=1
1530 NEXT I
1540 XM=-100000
1550 FOR I=1 TO NN
1560 SG=0
1570 FOR J=1 TO NN
1580 SG=SG+A(I,J)*X(J)
1590 NEXT J
1600 Z(I)=SG
1610 NEXT I
1620 XM=0
1630 FOR I=1 TO NN
1640 IF ABS(XM)<ABS(Z(I)) THEN XM=Z(I)
1650 NEXT I
1660 FOR I=1 TO NN
1670 X(I)=Z(I)/XM
1680 NEXT I
1690 IF ABS((Y1-XM)/XM)>D THEN 1710
1700 GOTO 1730
1710 Y1=XM
1720 GOTO 1550
1730 X3=0
1740 FOR I=1 TO NN
1750 IF ABS(X3)<ABS(X(I)) THEN X3=X(I)
1760 NEXT I
1770 FOR I=1 TO NN
1780 X(I)=X(I)/X3
1790 NEXT I
```

```
1800 RETURN
1810 END
3200 N9=N-1
3210 FORX=1TON
3220 DI=A(X,1)
3230 IFDI=0THENPRINT"MATRIX IS SINGULAR"
3240 FORY=1TON9
3250 Y9=Y+1
3260 A(X,Y)=A(X,Y9)/DI
3270 NEXTY
3280 A(X,N)=1/DI
3290 FORZ=1TON
3300 IFZ=XTHEN3370
3310 O=A(Z,1)
3320 FORY=1TON9
3330 Y9=Y+1
3340 A(Z,Y)=A(Z,Y9)-A(X,Y)*O
3350 NEXTY
3360 A(Z,N)=-A(X,N)*O
3370 NEXTZ
3380 NEXTX
3390 RETURN
6020 FOR I=1 TO NN
6030 FOR J=1 TO NN
6040 A(I,J)=0.0
6050 NEXT J
6060 FOR J=1 TO NN+1
6070 XX(I,J)=0.0
6080 NEXT J
6090 NEXT I
6100 PRINT "DISPLACEMENT POSNS OF SUPPRESSED DISPLTS - IN ASCENDING ORDER"
6110 FOR I=1 TO NF
6115 PRINT "POSITION";I
6120 INPUT NS(I)
6130 NEXT I
6140 PRINT "ELASTIC MODULUS"
6150 INPUT E
6160 PRINT "DENSITY"
6170 INPUT RH
6180 PRINT "ELEMENT DETAILS FROM LEFT TO RIGHT":PRINT
6190 FOR LE=1 TO LS
6195 PRINT"ELEMENT NO";LE
6200 PRINT "I NODE (LEFT)"
6210 INPUT I
6220 PRINT "J NODE (RIGHT)"
6230 INPUT J
6240 PRINT "2ND MOM OF AREA"
6250 INPUT SA
6260 PRINT "CROSS SECTIONAL AREA"
6270 INPUT CA
6280 PRINT "ELEMENTAL LENGTH"
6290 INPUT L
6295 PRINT:PRINT"ELEMENT NO";LE;"UNDER COMPUTATION":PRINT
6300 ST(1,1)=12/L↑3
6310 ST(1,2)=-6/L↑2
6320 ST(1,4)=ST(1,2)
6330 ST(2,1)=ST(1,2)
6340 ST(4,1)=ST(1,2)
6350 ST(3,3)=ST(1,1)
6360 ST(1,3)=-12/L↑3
6370 ST(3,1)=ST(1,3)
6380 ST(2,2)=4/L
6390 ST(4,4)=ST(2,2)
6400 ST(2,3)=6/L↑2
6410 ST(3,2)=ST(2,3)
6420 ST(2,4)=2/L
6430 ST(4,2)=ST(2,4)
6440 ST(3,4)=6/L↑2
6450 ST(4,3)=ST(3,4)
6460 CN=E*SA
6470 FOR II=1 TO 4
6480 FOR JJ=1 TO 4
6490 ST(II,JJ)=ST(II,JJ)*CN
6500 NEXT JJ
6510 NEXT II
```

```
6520 SM(1,1)=156
6530 SM(3,3)=156
6540 SM(1,2)=-22*L
6550 SM(2,1)=-22*L
6560 SM(1,3)=54
6570 SM(3,1)=54
6580 SM(1,4)=13*L
6590 SM(4,1)=13*L
6600 SM(2,2)=4*L↑2
6610 SM(4,4)=4*L↑2
6620 SM(2,3)=-13*L
6630 SM(3,2)=-13*L
6640 SM(2,4)=-3*L↑2
6650 SM(4,2)=-3*L↑2
6660 SM(3,4)=22*L
6670 SM(4,3)=22*L
6680 CN=RH*CA*L/420
6690 FOR  II=1 TO 4
6700 FOR JJ=1 TO 4
6710 SM(II,JJ)=SM(II,JJ)*CN
6720 NEXT JJ
6730 NEXT II
6731 I1=2*I-2
6732 J1=2*J-2
6740 FOR II=1 TO 2
6750 FOR JJ=1 TO 2
6760 MG=I1+II
6770 TR=I1+JJ
6780 AC=J1+II
6790 XK=J1+JJ
6800 A(MG,TR)=A(MG,TR)+ST(II,JJ)
6810 XX(MG,TR)=XX(MG,TR)+SM(II,JJ)
6820 A(AC,TR)=A(AC,TR)+ST(II+2,JJ)
6830 XX(AC,TR)=XX(AC,TR)+SM(II+2,JJ)
6840 A(MG,XK)=A(MG,XK)+ST(II,JJ+2)
6850 XX(MG,XK)=XX(MG,XK)+SM(II,JJ+2)
6860 A(AC,XK)=A(AC,XK)+ST(II+2,JJ+2)
6870 XX(AC,XK)=XX(AC,XK)+SM(II+2,JJ+2)
6880 NEXT JJ
6890 NEXT II
6900 NEXT LE
6910 MM=0
6920 FOR I=1 TO NF
6930 N1=NS(I)
6940 MM=MM+1
6950 N1=N1-MM+1
6960 N2=NN-MM
6970 IF N2<N1 THEN 7040
6980 FOR II=N1 TO N2
6990 FOR JJ=N1 TO N2
7000 A(II,JJ)=A(II+1,JJ+1)
7010 XX(II,JJ)=XX(II+1,JJ+1)
7020 NEXT JJ
7030 NEXT II
7040 IF N1=1 THEN 7130
7050 FOR II=1 TO N1-1
7060 FOR JJ=N1 TO N2
7070 A(II,JJ)=A(II,JJ+1)
7080 XX(II,JJ)=XX(II,JJ+1)
7090 A(JJ,II)=A(JJ+1,II)
7100 XX(JJ,II)=XX(JJ+1,II)
7110 NEXT JJ
7120 NEXT II
7130 NEXT I
7140 PRINT "NO OF CONC MASSES"
7150 INPUT NC
7160 IF NC=0 THEN 7300
7170 FOR II=1 TO NC
7180 PRINT "NODAL POSN OF MASSES"
7190 INPUT PO
7200 PRINT "VALUE OF MASS AND MASS MOM  OF INERTIA"
7210 INPUT MS,MI
7220 I1=2*PO-1
7225 I9=2*PO-1
7230 FOR JJ=1 TO NF
```

```
 7240 IF NS(JJ)<I9 THEN I1=I1-1
 7250 NEXT JJ
 7260 J1=I1+1
 7270 XX(I1,I1)=XX(I1,I1)+MS
 7280 XX(J1,J1)=XX(J1,J1)+MI
 7290 NEXT II
 7300 PRINT:PRINT"THE STIFFNESS MATRIX IS BEING INVERTED":PRINT
 7310 GOSUB 3200
 7320 FOR I=1 TO N
 7330 FOR J=1 TO N
 7340 VE(J)=0
 7350 FOR K=1 TO N
 7360 VE(J)=VE(J)+A(I,K)*XX(K,J)
 7370 NEXT K
 7380 NEXT J
 7385 FOR J=1 TO N
 7390 A(I,J)=VE(J)
 7395 NEXT J
 7400 NEXT I
 7410 D=0.001
 7420 GOTO 200
READY.
```

APPENDIX 7

```
10 PRINT:PRINT:PRINT"VIBRATION OF RIGID JOINTED PLANE FRAMES":PRINT:PRINT
20 PRINT "NO OF NODAL POINTS"
30 INPUT NN
40 PRINT "NO OF ELEMENTS"
50 INPUT MS
60 PRINT "NO OF SUPPRESSED DISPLTS"
70 INPUT NF
80 N3=NN*3
90 NP=N3+1
100 N=N3-NF
110 PRINT "NO OF FREQUENCIES-THIS MUST BE <=";N
120 INPUT M1
130 DIM A(N3,N3),XX(N3,NP),AM(M1),VC(N,M1),X(60),IG(N),Z(N),NS(NF),Y(NN)
140 DIM DC(6,6),SM(6,6),S2(6,6),VE(N)
150 GOTO 6020
200 PRINT:PRINT"THE EIGENVALUES ARE BEING DETERMINED":PRINT
210 GOSUB 500
220 FOR I=1 TO M1
240 PRINT "EIGENVALUE=";AM(I)
250 PRINT "FREQUENCY=";SQR(1/AM(I))/(2*π)
260 PRINT "EIGENVECTOR IS"
270 PRINT "    "
280 FOR J=1 TO N
300 PRINT VC(J,I);" ";
320 NEXT J
325 PRINT:PRINT
330 PRINT"TO CONTINUE, PRESS ANY KEY":PRINT
335 GETA$:IFA$="" THEN GOTO 335
340 NEXT I
360 END
500 MN=N
510 NN=N
520 GOSUB 1500
530 M=1
540 FOR I=1 TO NN
550 VC(I,M)=X(I)
560 XX(I,M)=VC(I,M)
570 NEXT I
580 AM(M)=XM
590 IF M1<2 THEN 220
600 FOR M=2 TO M1
610 FOR I=1 TO NN
620 K4=ABS(XX(I,M-1)-1)
630 IF K4<0.00001 THEN IR=I
650 NEXT I
660 IG(M-1)=IR
670 FOR I=1 TO NN
680 XX(MN-I+1,MN-M+3)=A(IR,I)
690 NEXT I
700 FOR I=1 TO NN
710 FOR J=1 TO NN
720 Z1=MN-J+1
730 Z2=MN-M+3
740 A(I,J)=A(I,J)-XX(I,M-1)*XX(Z1,Z2)
750 NEXT J
760 NEXT I
770 FOR I=1 TO NN
780 IF I=IR THEN 910
790 IF I>IR THEN K1=I-1
810 IF I<=IR THEN K1=I
830 FOR J=1 TO NN
840 IF J=IR THEN 900
850 IF J>IR THEN K2=J-1
870 IF J<=IR THEN K2=J
890 A(K1,K2)=A(I,J)
900 NEXT J
910 NEXT I
920 NN=NN-1
930 M3=NN
940 IF M<>MN THEN 980
950 XM=A(1,1)
960 X(1)=1
```

```
970 GOTO 990
980 GOSUB 1500
990 FOR I=1 TO NN
1000 XX(I,M)=X(I)
1010 NEXT I
1020 AM(M)=XM
1030 M4=M-1
1040 M5=1000-M4
1050 FOR M8=M5 TO 999
1060 M6=M3+1
1070 M2=1000-M8
1080 M7=IG(M2)+1
1090 IF M6<M7 THEN 1150
1100 N9=1000-M7
1110 N8=1000-M6
1120 FOR I3=N8 TO N9
1125 I=1000-I3
1130 X(I)=X(I-1)
1140 NEXT I3
1150 J=IG(M2)
1160 X(J)=0
1170 SM=0
1180 FOR I=1 TO M6
1190 Z3=MN-I+1
1200 Z4=MN-M2+2
1210 SM=SM+XX(Z3,Z4)*X(I)
1220 NEXT I
1230 XK=(AM(M2)-XM)/SM
1240 FOR I=1 TO M6
1250 X(I)=XX(I,M2)-XK*X(I)
1260 NEXT I
1270 SM=0
1280 FOR I=1 TO M6
1290 IF ABS(SM)<ABS(X(I)) THEN SM=X(I)
1310 NEXT I
1320 FOR I=1 TO M6
1330 X(I)=X(I)/SM
1340 NEXT I
1350 M3=M3+1
1360 IF M2<>1 THEN 1400
1370 FOR I=1 TO M3
1380 VC(I,M)=X(I)
1390 NEXT I
1400 NEXT M8
1405 NEXT M
1410 RETURN
1420 END
1500 Y1=100000
1510 FOR I=1 TO NN
1520 X(I)=1
1530 NEXT I
1540 XM=-100000
1550 FOR I=1 TO NN
1560 SG=0
1570 FOR J=1 TO NN
1580 SG=SG+A(I,J)*X(J)
1590 NEXT J
1600 Z(I)=SG
1610 NEXT I
1620 XM=0
1630 FOR I=1 TO NN
1640 IF ABS(XM)<ABS(Z(I)) THEN XM=Z(I)
1650 NEXT I
1660 FOR I=1 TO NN
1670 X(I)=Z(I)/XM
1680 NEXT I
1690 IF ABS((Y1-XM)/XM)>D THEN 1710
1700 GOTO 1730
1710 Y1=XM
1720 GOTO 1550
1730 X3=0
1740 FOR I=1 TO NN
1750 IF ABS(X3)<ABS(X(I)) THEN X3=X(I)
1760 NEXT I
1770 FOR I=1 TO NN
```

```
1780 X(I)=X(I)/X3
1790 NEXT I
1800 RETURN
1810 END
3200 N9=N-1
3210 FORX=1TON
3220 DI=A(X,1)
3230 IFDI=0THENPRINT"MATRIX IS SINGULAR"
3240 FORY=1TON9
3250 Y9=Y+1
3260 A(X,Y)=A(X,Y9)/DI
3270 NEXTY
3280 A(X,N)=1/DI
3290 FORZ=1TON
3300 IFZ=XTHEN3370
3310 O=A(Z,1)
3320 FORY=1TON9
3330 Y9=Y+1
3340 A(Z,Y)=A(Z,Y9)-A(X,Y)*O
3350 NEXTY
3360 A(Z,N)=-A(X,N)*O
3370 NEXTZ
3380 NEXTX
3390 RETURN
6020 FOR I=1 TO N3
6030 FOR J=1 TO N3
6040 A(I,J)=0.0
6050 NEXT J
6060 FOR J=1 TO NP
6070 XX(I,J)=0.0
6080 NEXT J
6090 NEXT I
6100 PRINT "FEED IN THE NODAL COORDS"
6110 FOR I=1 TO NN
6120 PRINT "X COORD OF NODE";I
6130 INPUT X(I)
6140 PRINT "Y COORD OF NODE";I
6150 INPUT Y(I)
6160 NEXT I
6170 PRINT "FEED DISPLACEMENT POSNS OF SUPPRESSED DISPLTS-IN ASCENDING ORDER"
6180 FOR I=1 TO NF
6190 PRINT "POSITION";I
6200 INPUT NS(I)
6210 NEXT I
6220 PRINT "ELASTIC MODULUS"
6230 INPUT E
6240 PRINT "DENSITY"
6250 INPUT RH
6260 PRINT "MEMBER DETAILS"
6270 FOR ME=1 TO MS
6275 PRINT"MEMBER NO";ME
6280 PRINT "I NODE"
6290 INPUT I
6300 PRINT "J NODE"
6310 INPUT J
6320 PRINT "2ND MOM OF AREA"
6330 INPUT SA
6340 PRINT "CROSS SECTIONAL AREA"
6350 INPUT CA
6355 PRINT:PRINT"MEMBER NO";ME;"UNDER COMPUTATION":PRINT
6360 L=SQR((X(J)-X(I))↑2+(Y(J)-Y(I))↑2)
6370 C=(X(J)-X(I))/L
6380 S=(Y(J)-Y(I))/L
6390 FOR II=1 TO 6
6400 FOR JJ=1 TO 6
6410 DC(II,JJ)=0
6420 SM(II,JJ)=0
6430 NEXT JJ
6440 NEXT II
6450 DC(1,1)=C
6460 DC(2,2)=C
6470 DC(4,4)=C
6480 DC(5,5)=C
6490 DC(1,2)=S
6500 DC(4,5)=S
```

```
6510 DC(2,1)=-S
6520 DC(5,4)=-S
6530 DC(3,3)=1
6540 DC(6,6)=1
6600 I3=3*I
6610 I2=I3-1
6620 I1=I3-2
6630 J3=3*J
6640 J2=J3-1
6650 J1=J3-2
6660 C1=12*E*SA*S↑2/L↑3+C↑2*CA*E/L
6670 C2=12*E*SA*C*S/L↑3-C*S*CA*E/L
6680 C3=12*E*SA*C↑2/L↑3+S↑2*CA*E/L
6690 C4=6*E*SA*S/L↑2
6700 C5=6*E*SA*C/L↑2
6710 C6=4*E*SA/L
6720 A(I1,I1)=A(I1,I1)+C1
6730 A(I2,I1)=A(I2,I1)-C2
6740 A(I1,I2)=A(I1,I2)-C2
6750 A(I3,I1)=A(I3,I1)+C4
6760 A(I1,I3)=A(I1,I3)+C4
6770 A(J1,I1)=A(J1,I1)-C1
6780 A(I1,J1)=A(I1,J1)-C1
6790 A(J2,I1)=A(J2,I1)+C2
6800 A(I1,J2)=A(I1,J2)+C2
6810 A(J3,I1)=A(J3,I1)+C4
6820 A(I1,J3)=A(I1,J3)+C4
6830 A(I2,I2)=A(I2,I2)+C3
6840 A(I3,I2)=A(I3,I2)-C5
6850 A(I2,I3)=A(I2,I3)-C5
6860 A(J1,I2)=A(J1,I2)+C2
6870 A(I2,J1)=A(I2,J1)+C2
6880 A(J2,I2)=A(J2,I2)-C3
6890 A(I2,J2)=A(I2,J2)-C3
6900 A(J3,I2)=A(J3,I2)-C5
6910 A(I2,J3)=A(I2,J3)-C5
6920 A(I3,I3)=A(I3,I3)+C6
6930 A(J1,I3)=A(J1,I3)-C4
6940 A(I3,J1)=A(I3,J1)-C4
6950 A(J2,I3)=A(J2,I3)+C5
6960 A(I3,J2)=A(I3,J2)+C5
6970 A(J3,I3)=A(J3,I3)+0.5*C6
6980 A(I3,J3)=A(I3,J3)+0.5*C6
6990 A(J1,J1)=A(J1,J1)+C1
7000 A(J2,J1)=A(J2,J1)-C2
7010 A(J1,J2)=A(J1,J2)-C2
7020 A(J3,J1)=A(J3,J1)-C4
7030 A(J1,J3)=A(J1,J3)-C4
7040 A(J2,J2)=A(J2,J2)+C3
7050 A(J3,J2)=A(J3,J2)+C5
7060 A(J2,J3)=A(J2,J3)+C5
7070 A(J3,J3)=A(J3,J3)+C6
7080 C1=RH*CA*L/6
7090 C2=RH*CA*L/420
7100 SM(1,1)=2*C1
7110 SM(4,1)=C1
7120 SM(2,2)=156*C2
7130 SM(3,2)=-22*L*C2
7140 SM(5,2)=54*C2
7150 SM(6,2)=13*L*C2
7160 SM(2,3)=-22*L*C2
7170 SM(3,3)=4*L↑2*C2
7180 SM(5,3)=-13*L*C2
7190 SM(6,3)=-3*L↑2*C2
7200 SM(1,4)=C1
7210 SM(4,4)=2*C1
7220 SM(2,5)=54*C2
7230 SM(3,5)=-13*L*C2
7240 SM(5,5)=156*C2
7250 SM(6,5)=22*L*C2
7260 SM(2,6)=13*L*C2
7270 SM(3,6)=-3*L↑2*C2
7280 SM(5,6)=22*L*C2
7290 SM(6,6)=4*L↑2*C2
7300 FOR II=1 TO 6
```

```
7310 FOR JJ=1 TO 6
7320 S2(II,JJ)=0
7321 NEXT JJ
7322 NEXT II
7323 FOR II=1 TO 6
7324 FOR JJ=1 TO 6
7330 FOR KK=1 TO 6
7340 S2(II,JJ)=S2(II,JJ)+SM(II,KK)*DC(KK,JJ)
7350 NEXT KK
7360 NEXT JJ
7370 NEXT II
7380 FOR II=1 TO 6
7390 FOR JJ=1 TO 6
7400 SM(II,JJ)=0
7401 NEXT JJ
7402 NEXT II
7403 FOR II=1 TO 6
7404 FOR JJ=1 TO 6
7410 FOR KK=1 TO 6
7420 SM(II,JJ)=SM(II,JJ)+DC(KK,II)*S2(KK,JJ)
7430 NEXT KK
7440 NEXT JJ
7450 NEXT II
7460 I1=3*I-3
7470 J1=3*J-3
7480 FOR II=1 TO 3
7490 FOR JJ=1 TO 3
7500 MG=I1+II
7510 TR=I1+JJ
7520 AC=J1+II
7530 XK=J1+JJ
7540 XX(MG,TR)=XX(MG,TR)+SM(II,JJ)
7550 XX(AC,TR)=XX(AC,TR)+SM(II+3,JJ)
7560 XX(MG,XK)=XX(MG,XK)+SM(II,JJ+3)
7570 XX(AC,XK)=XX(AC,XK)+SM(II+3,JJ+3)
7580 NEXT JJ
7590 NEXT II
7600 NEXT ME
7610 PRINT "NO OF CONCENTRATED MASSES"
7620 INPUT NC
7630 IF NC=0 THEN 7780
7640 FOR II=1 TO NC
7650 PRINT "NODAL POSN OF MASS"
7660 INPUT PO
7670 PRINT "VALUE OF MASS"
7680 INPUT MS
7690 PRINT "VALUE OF MASS MOM OF INERTIA"
7700 INPUT MI
7710 K1=3*PO
7720 J1=K1-1
7730 I1=K1-2
7740 XX(I1,I1)=XX(I1,I1)+MS
7750 XX(J1,J1)=XX(J1,J1)+MS
7760 XX(K1,K1)=XX(K1,K1)+MI
7770 NEXT II
7780 MM=0
7790 FOR I=1 TO NF
7800 N1=NS(I)
7810 MM=MM+1
7820 N1=N1-MM+1
7830 N2=N3-MM
7840 IF N2<N1 THEN 7910
7850 FOR II=N1 TO N2
7860 FOR JJ=N1 TO N2
7870 A(II,JJ)=A(II+1,JJ+1)
7880 XX(II,JJ)=XX(II+1,JJ+1)
7890 NEXT JJ
7900 NEXT II
7910 IF N1=1 THEN 8020
7920 FOR II=1 TO N1-1
7930 FOR JJ=N1 TO N2
7940 A(II,JJ)=A(II,JJ+1)
7950 XX(II,JJ)=XX(II,JJ+1)
7960 A(JJ,II)=A(JJ+1,II)
7970 XX(JJ,II)=XX(JJ+1,II)
```

```
 7980 NEXT JJ
 7990 NEXT II
 8020 NEXT I
 8030 PRINT:PRINT"THE STIFFNESS MATRIX IS BEING INVERTED":PRINT
 8040 GOSUB 3200
 8050 FOR I=1 TO N
 8060 FOR J=1 TO N
 8070 VE(J)=0
 8080 FOR II=1 TO N
 8090 VE(J)=VE(J)+A(I,II)*XX(II,J)
 8100 NEXT II
 8110 NEXT J
 8120 FOR J=1 TO N
 8130 A(I,J)=VE(J)
 8140 NEXT J
 8150 NEXT I
 8160 D=0.001
 8170 GOTO 200
 8180 END
READY.
```

APPENDIX 8

```
8 PRINT:PRINT:PRINT "FREE VIBRATION OF PIN JOINTED SPACE TRUSSES":PRINT:PRINT
10 PRINT "NO OF PIN JOINTS"
20 INPUT NJ
30 PRINT "NO OF SUPPRESSED DISPLTS"
40 INPUT NF
50 N3=3*NJ
60 N=N3-NF
70 NP=N3+1
80 PRINT "NO OF MEMBERS"
90 INPUT MS
100 PRINT "NO OF FREQUENCIES REQD-THIS MUST BE <=";N
110 INPUT M1
120 DIM A(N3,N3),XX(N3,NP),VC(N,M1),AM(M1),VE(N),NS(NF)
130 DIM X(60),Y(NJ),Z(60),SK(6,6),SM(6,6),IG(N)
150 GOTO 6020
200 PRINT:PRINT"THE EIGENVALUES ARE BEING DETERMINED":PRINT
210 GOSUB 500
220 FOR I=1 TO M1
240 PRINT "EIGENVALUE=";AM(I)
250 PRINT "FREQUENCY=";SQR(1/AM(I))/(2*π)
260 PRINT "EIGENVECTOR IS"
270 PRINT "   "
280 FOR J=1 TO N
300 PRINT VC(J,I);" ";
320 NEXT J
325 PRINT:PRINT:PRINT"TO CONTINUE, PRESS ANY KEY":PRINT
330 GET A$: IF A$="" THEN 330
340 NEXT I
360 END
500 MN=N
510 NN=N
520 GOSUB 1500
530 M=1
540 FOR I=1 TO NN
550 VC(I,M)=X(I)
560 XX(I,M)=VC(I,M)
570 NEXT I
580 AM(M)=XM
590 IF M1<2 THEN 220
600 FOR M=2 TO M1
610 FOR I=1 TO NN
620 K4=ABS(XX(I,M-1)-1)
630 IF K4<0.00001 THEN IR=I
650 NEXT I
660 IG(M-1)=IR
670 FOR I=1 TO NN
680 XX(MN-I+1,MN-M+3)=A(IR,I)
690 NEXT I
700 FOR I=1 TO NN
710 FOR J=1 TO NN
720 Z1=MN-J+1
730 Z2=MN-M+3
740 A(I,J)=A(I,J)-XX(I,M-1)*XX(Z1,Z2)
750 NEXT J
760 NEXT I
770 FOR I=1 TO NN
780 IF I=IR THEN 910
790 IF I>IR THEN K1=I-1
810 IF I<=IR THEN K1=I
830 FOR J=1 TO NN
840 IF J=IR THEN 900
850 IF J>IR THEN K2=J-1
870 IF J<=IR THEN K2=J
890 A(K1,K2)=A(I,J)
900 NEXT J
910 NEXT I
920 NN=NN-1
930 M3=NN
940 IF M<>MN THEN 980
950 XM=A(1,1)
960 X(1)=1
970 GOTO 990
```

```
980 GOSUB 1500
990 FOR I=1 TO NN
1000 XX(I,M)=X(I)
1010 NEXT I
1020 AM(M)=XM
1030 M4=M-1
1040 M5=1000-M4
1050 FOR M8=M5 TO 999
1060 M6=M3+1
1070 M2=1000-M8
1080 M7=IG(M2)+1
1090 IF M6<M7 THEN 1150
1100 N9=1000-M7
1110 N8=1000-M6
1120 FOR I3=N8 TO N9
1125 I=1000-I3
1130 X(I)=X(I-1)
1140 NEXT I3
1150 J=IG(M2)
1160 X(J)=0
1170 SM=0
1180 FOR I=1 TO M6
1190 Z3=MN-I+1
1200 Z4=MN-M2+2
1210 SM=SM+XX(Z3,Z4)*X(I)
1220 NEXT I
1230 XK=(AM(M2)-XM)/SM
1240 FOR I=1 TO M6
1250 X(I)=XX(I,M2)-XK*X(I)
1260 NEXT I
1270 SM=0
1280 FOR I=1 TO M6
1290 IF ABS(SM)<ABS(X(I)) THEN SM=X(I)
1310 NEXT I
1320 FOR I=1 TO M6
1330 X(I)=X(I)/SM
1340 NEXT I
1350 M3=M3+1
1360 IF M2<>1 THEN 1400
1370 FOR I=1 TO M3
1380 VC(I,M)=X(I)
1390 NEXT I
1400 NEXT M8
1405 NEXT M
1410 RETURN
1420 END
1500 Y1=100000
1510 FOR I=1 TO NN
1520 X(I)=1
1530 NEXT I
1540 XM=-100000
1550 FOR I=1 TO NN
1560 SG=0
1570 FOR J=1 TO NN
1580 SG=SG+A(I,J)*X(J)
1590 NEXT J
1600 Z(I)=SG
1610 NEXT I
1620 XM=0
1630 FOR I=1 TO NN
1640 IF ABS(XM)<ABS(Z(I)) THEN XM=Z(I)
1650 NEXT I
1660 FOR I=1 TO NN
1670 X(I)=Z(I)/XM
1680 NEXT I
1690 IF ABS((Y1-XM)/XM)>D THEN 1710
1700 GOTO 1730
1710 Y1=XM
1720 GOTO 1550
1730 X3=0
1740 FOR I=1 TO NN
1750 IF ABS(X3)<ABS(X(I)) THEN X3=X(I)
1760 NEXT I
1770 FOR I=1 TO NN
1780 X(I)=X(I)/X3
```

```
1790 NEXT I
1800 RETURN
1810 END
3200 N9=N-1
3210 FORX=1TON
3220 DI=A(X,1)
3230 IFDI=0THENPRINT"MATRIX IS SINGULAR"
3240 FORY=1TON9
3250 Y9=Y+1
3260 A(X,Y)=A(X,Y9)/DI
3270 NEXTY
3280 A(X,N)=1/DI
3290 FORZ=1TON
3300 IFZ=XTHEN3370
3310 O=A(Z,1)
3320 FORY=1TON9
3330 Y9=Y+1
3340 A(Z,Y)=A(Z,Y9)-A(X,Y)*O
3350 NEXTY
3360 A(Z,N)=-A(X,N)*O
3370 NEXTZ
3380 NEXTX
3390 RETURN
6020 FOR I=1 TO N3
6030 FOR J=1 TO N3
6040 A(I,J)=0
6050 NEXT J
6060 FOR J=1 TO NP
6070 XX(I,J)=0
6080 NEXT J
6090 NEXT I
6100 PRINT "NODAL COORDINATES"
6110 FOR I=1 TO NJ
6120 PRINT "X COORD FOR NODE";I
6130 INPUT X(I)
6140 PRINT "Y COORD FOR NODE";I
6150 INPUT Y(I)
6160 PRINT "Z COORD FOR NODE";I
6170 INPUT Z(I)
6180 NEXT I
6190 PRINT "MEMBER DETAILS"
6200 FOR ME =1 TO MS
6205 PRINT"MEMBER NO";ME
6210 PRINT "I NODE"
6220 INPUT I
6230 PRINT "J NODE "
6240 INPUT J
6250 PRINT "CROSS SECTIONAL AREA"
6260 INPUT A
6270 PRINT "ELASTIC MODULUS"
6280 INPUT E
6290 PRINT "DENSITY"
6300 INPUT RH
6305 PRINT:PRINT"MEMBER NO";ME;"UNDER COMPUTATION":PRINT
6310 FOR II=1 TO 6
6320 FOR JJ=1 TO 6
6330 SK(II,JJ)=0
6340 SM(II,JJ)=0
6350 NEXT JJ
6360 NEXT II
6370 L=SQR((X(J)-X(I))↑2+(Y(J)-Y(I))↑2+(Z(J)-Z(I))↑2)
6380 XC=(X(J)-X(I))/L
6390 YC=(Y(J)-Y(I))/L
6400 ZC=(Z(J)-Z(I))/L
6410 SK(1,1)=XC↑2
6420 SK(2,1)=XC*YC
6430 SK(2,2)=YC↑2
6440 SK(3,1)=XC*ZC
6450 SK(3,2)=YC*ZC
6460 SK(3,3)=ZC↑2
6470 SK(4,1)=-XC↑2
6480 SK(4,2)=-XC*YC
6490 SK(4,3)=-XC*ZC
6500 SK(5,1)=-XC*YC
6510 SK(5,2)=-YC↑2
```

```
6520 SK(5,3)=-YC*ZC
6530 SK(6,1)=-XC*ZC
6540 SK(6,2)=-YC*ZC
6550 SK(6,3)=-ZC↑2
6560 FOR II=1 TO 2
6570 FOR JJ=II+1 TO 3
6580 SK(II,JJ)=SK(JJ,II)
6590 NEXT JJ
6600 NEXT II
6610 FOR II=1 TO 3
6620 FOR JJ=1 TO 3
6630 SK(II+3,JJ+3)=SK(II,JJ)
6640 SK(II,JJ+3)=SK(JJ+3,II)
6650 NEXT JJ
6660 NEXT II
6670 CN=RH*A*L
6680 FOR II=1 TO 6
6690 SM(II,II)=CN/3
6700 NEXT II
6710 FOR II=1 TO 3
6720 SM(II+3,II)=CN/6
6730 SM(II,II+3)=CN/6
6740 NEXT II
6750 FOR II=1 TO 6
6760 FOR JJ=1 TO 6
6770 SK(II,JJ)=SK(II,JJ)*A*E/L
6780 NEXT JJ
6790 NEXT II
6800 I1=3*I-3
6810 J1=3*J-3
6820 FOR II=1 TO 3
6830 FOR JJ=1 TO 3
6840 MM=I1+II
6850 MN=I1+JJ
6860 NM=J1+II
6870 NN=J1+JJ
6880 A(MM,MN)=A(MM,MN)+SK(II,JJ)
6890 A(MM,NN)=A(MM,NN)+SK(II,JJ+3)
6900 A(NM,MN)=A(NM,MN)+SK(II+3,JJ)
6910 A(NM,NN)=A(NM,NN)+SK(II+3,JJ+3)
6920 XX(MM,MN)=XX(MM,MN)+SM(II,JJ)
6930 XX(MM,NN)=XX(MM,NN)+SM(II,JJ+3)
6940 XX(NM,MN)=XX(NM,MN)+SM(II+3,JJ)
6950 XX(NM,NN)=XX(NM,NN)+SM(II+3,JJ+3)
6960 NEXT JJ
6970 NEXT II
6980 NEXT ME
6990 PRINT "FEED DISPLACEMENT POSNS OF SUPPRESSED DISPLTS-IN ASCENDING ORDER"
7000 FOR II=1 TO NF
7005 PRINT "POSITION";II
7010 INPUT NS(II)
7020 NEXT II
7030 PRINT "NO OF CONCENTRATED MASSES"
7040 INPUT NC
7045 IF NC=0 THEN 7170
7050 FOR II=1 TO NC
7060 PRINT "NODAL POSITION OF MASS"
7070 INPUT PO
7080 PRINT "VALUE OF MASS"
7090 INPUT MC
7100 K1=3*PO
7110 J1=K1-1
7120 I1=K1-2
7130 XX(I1,I1)=XX(I1,Ii)+MC
7140 XX(J1,J1)=XX(J1,J1)+MC
7150 XX(K1,K1)=XX(K1,K1)+MC
7160 NEXT II
7170 MM=0
7180 FOR I=1 TO NF
7190 N1=NS(I)
7200 MM=MM+1
7210 N1=N1-MM+1
7220 N2=N3-MM
7230 IF N2<N1 THEN 7300
7240 FOR II=N1 TO N2
```

```
7250 FOR JJ=N1 TO N2
7260 A(II,JJ)=A(II+1,JJ+1)
7270 XX(II,JJ)=XX(II+1,JJ+1)
7280 NEXT JJ
7290 NEXT II
7300 IF N1=1 THEN 7390
7310 FOR II=1 TO N1-1
7320 FOR JJ=N1 TO N2
7330 A(II,JJ)=A(II,JJ+1)
7340 A(JJ,II)=A(JJ+1,II)
7350 XX(II,JJ)=XX(II,JJ+1)
7360 XX(JJ,II)=XX(JJ+1,II)
7370 NEXT JJ
7380 NEXT II
7390 NEXT I
7400 PRINT:PRINT"THE STIFFNESS MATRIX IS BEING INVERTED":PRINT
7410 GOSUB 3200
7420 FOR I=1 TO N
7430 FOR J=1 TO N
7440 VE(J)=0
7450 FOR K=1 TO N
7460 VE(J)=VE(J)+A(I,K)*XX(K,J)
7470 NEXT K
7480 NEXT J
7490 FOR J=1 TO N
7500 A(I,J)=VE(J)
7510 NEXT J
7520 NEXT I
7530 D=0.001
7540 GOTO 200
READY.
```

APPENDIX 9

```
140 PRINT:PRINT"STATIC ANALYSIS OF RIGID JOINTED SPACE FRAMES":PRINT
150 PRINT"FEED NO OF ACTUAL NODES":INPUTNJ
160 PRINT"FEED TOTAL NO OF SUPPRESSED DISPLACEMENTS":PRINT
170 INPUTNF
180 PRINT"FEED NO OF IMAGINARY NODES"
190 INPUTIM
200 N6=6*NJ:NN=NJ+IM
210 PRINT"FEED NO OF DIFFERENT TYPES OF MEMBER"
220 INPUTML
230 PRINT"FEED NO OF MEMBERS"
240 INPUTMS
250 M3=3*MS
260 IFML=1THENMZ=1
270 IFML>1THENMZ=MS
280 DIMDC(3,3),DT(3,3),SK(12,12),SX(12,12),UL(12),UG(12)
290 DIMBM(4,12),SR(4)
300 DIME(ML),G(ML),AR(ML),IX(ML),IY(ML),IZ(ML),NM(MZ)
310 DIMX(NN),Y(NN),Z(NN),IJ(M3)
320 DIMU(N6),NS(NF)
330 DIMXA(3),YA(3),ZA(3)
340 DIMS2(3),S3(3),S4(3),Z1(3),Z2(3),Z3(3),ZP(3,3)
350 FORII=1TONN
360 PRINT"X COORD FOR NODE";II
370 INPUTX(II)
380 PRINT"Y COORD FOR NODE";II
390 INPUTY(II)
400 PRINT"Z COORD FOR NODE";II
410 INPUTZ(II)
420 NEXTII
430 FORII=1TOML
440 PRINT"E FOR MEMBER TYPE";II
450 INPUTE(II)
460 PRINT"G FOR MEMBER TYPE";II
470 INPUTG(II)
480 PRINT"CROSS SECTNL AREA FOR MEMBER TYPE";II
490 INPUTAR(II)
500 PRINT"TORSIONAL CONSTANT FOR MEMBER TYPE";II
510 INPUTIX(II)
520 PRINT"2ND MOA ABOUT XY PLANE FOR MEMBER TYPE";II
530 INPUTIY(II)
540 PRINT"2ND MOA ABOUT XZ PLANE FOR MEMBER TYPE";;II
550 INPUTIZ(II)
560 NEXTII
570 MX=0
580 FORME=1TOMS
590 PRINT"I NODE FOR MEMBER";ME
600 INPUTI
610 PRINT"J NODE FOR MEMBER";ME
620 INPUTJ
630 PRINT"K NODE FOR MEMBER";ME
640 INPUTK
650 M8=3*ME
660 IJ(M8-2)=I
670 IJ(M8-1)=J
680 IJ(M8)=K
690 IFABS(J-I)>MXTHENMX=ABS(J-I)
700 IFML=1THENNM(1)=1
710 IFML=1THEN740
720 PRINT"FEED MEMBER TYPE FOR MEMBER";ME
730 INPUTNM(ME)
740 NEXTME
750 NW=(MX+1)*6
760 NT=NW+N6
770 DIMA(NT,NW),Q(NT),C(NT)
780 PRINT:PRINT"FEED THE NUMBER OF NODES WITH CONCENTRATED LOADS":PRINT
790 INPUTNC
800 FORII=1TONC
810 PRINT"FEED NODAL POSITION";II
820 INPUTPW
830 PRINT"FEED VALUE OF FORCE IN X0 DIRN"
840 INPUTLW
850 Q(PW*6-5)=LW
```

```
860 PRINT"FEED VALUE OF FORCE IN Y0 DIRN"
870 INPUTLW
880 Q(PW*6-4)=LW
890 PRINT"FEED VALUE OF FORCE IN Z0 DIRN"
900 INPUTLW
910 Q(PW*6-3)=LW
920 PRINT"FEED VALUE OF COUPLE IN X0 DIRN"
930 INPUTLW
940 Q(PW*6-2)=LW
950 PRINT"FEED VALUE OF COUPLE IN Y0 DIRN"
960 INPUTLW
970 Q(PW*6-1)=LW
980 PRINT"FEED VALUE OF COUPLE IN Z0 DIRN"
990 INPUTLW
1000 Q(PW*6)=LW
1010 NEXTII
1020 PRINT:PRINT"FEED (POSITIONS) OF SUPPRESSED DISPLACEMENTS":PRINT
1030 FORII=1TONF
1040 PRINT"DISPLACEMENT POSITION";II
1050 INPUTN7
1060 NS(II)=N7
1070 NEXTII
1080 FORME=1TOMS
1090 FORII=1TO12
1100 FORJJ=1TO12
1110 SK(II,JJ)=0
1120 NEXTJJ,II
1130 PRINT"MEMBER NO";ME;"UNDER COMPUTATION"
1140 M8=3*ME
1150 I=IJ(M8-2)
1160 J=IJ(M8-1)
1170 K=IJ(M8)
1180 IFML>1THEN1260
1190 E=E(1)
1200 G=G(1)
1210 A=AR(1)
1220 IX=IX(1)
1230 IY=IY(1)
1240 IZ=IZ(1)
1250 GOTO1320
1260 E=E(NM(ME))
1270 G=G(NM(ME))
1280 A=AR(NM(ME))
1290 IX=IX(NM(ME))
1300 IY=IY(NM(ME))
1310 IZ=IZ(NM(ME))
1320 LE=SQR((X(J)-X(I))↑2+(Y(J)-Y(I))↑2+(Z(J)-Z(I))↑2)
1330 SK(1,1)=A*E/LE
1340 SK(7,7)=A*E/LE
1350 SK(7,1)=-SK(1,1)
1360 SK(1,7)=-SK(1,1)
1370 SK(2,2)=12*E*IZ/LE↑3
1380 SK(8,8)=SK(2,2)
1390 SK(8,2)=-SK(2,2)
1400 SK(2,8)=-SK(2,2)
1410 SK(3,3)=12*E*IY/LE↑3
1420 SK(9,9)=SK(3,3)
1430 SK(9,3)=-SK(3,3)
1440 SK(3,9)=-SK(3,3)
1450 SK(4,4)=G*IX/LE
1460 SK(10,10)=SK(4,4)
1470 SK(10,4)=-SK(4,4)
1480 SK(4,10)=-SK(4,4)
1490 SK(5,5)=4*E*IY/LE
1500 SK(11,11)=SK(5,5)
1510 SK(11,5)=SK(5,5)/2
1520 SK(5,11)=SK(11,5)
1530 SK(6,6)=4*E*IZ/LE
1540 SK(12,12)=SK(6,6)
1550 SK(12,6)=SK(6,6)/2
1560 SK(6,12)=SK(6,6)/2
1570 SK(6,2)=6*E*IZ/LE↑2
1580 SK(2,6)=SK(6,2)
1590 SK(12,2)=SK(6,2)
1600 SK(2,12)=SK(6,2)
```

```
1610 SK(8,6)=-SK(6,2)
1620 SK(6,8)=-SK(6,2)
1630 SK(12,8)=-SK(6,2)
1640 SK(8,12)=-SK(6,2)
1650 SK(9,5)=6*E*IY/LE↑2
1660 SK(5,9)=SK(9,5)
1670 SK(11,9)=SK(9,5)
1680 SK(9,11)=SK(9,5)
1690 SK(5,3)=-SK(9,5)
1700 SK(3,5)=-SK(9,5)
1710 SK(11,3)=-SK(9,5)
1720 SK(3,11)=-SK(9,5)
1730 GOSUB2380
1740 FORLL=1TO4
1750 FORII=1TO3
1760 I9=LL*3-3+II
1770 FORMM=1TO4
1780 FORJJ=1TO3
1790 J9=MM*3-3+JJ
1800 SX(I9,J9)=0
1810 FORKK=1TO3
1820 K9=3*MM-3+KK
1830 SX(I9,J9)=SX(I9,J9)+SK(I9,K9)*DC(KK,JJ)
1840 NEXTKK
1850 NEXTJJ
1860 NEXTMM
1870 NEXTII
1880 NEXTLL
1890 FORLL=1TO4
1900 FORII=1TO3
1910 I9=LL*3-3+II
1920 FORMM=1TO4
1930 FORJJ=1TO3
1940 J9=MM*3-3+JJ
1950 SK(I9,J9)=0
1960 FORKK=1TO3
1970 K9=3*LL-3+KK
1980 SK(I9,J9)=SK(I9,J9)+DT(II,KK)*SX(K9,J9)
1990 NEXTKK
2000 NEXTJJ
2010 NEXTMM
2020 NEXTII
2030 NEXTLL
2040 I3=6*I-6
2050 J3=6*J-6
2060 FORJJ=1TO2
2070 NR=J3
2080 IFJJ=1THENNR=I3
2090 FORJ9=1TO6
2100 NR=NR+1
2110 II=(JJ-1)*6+J9
2120 FORKK=1TO2
2130 N9=J3
2140 IFKK=1THENN9=I3
2150 FORK9=1TO6
2160 LL=(KK-1)*6+K9
2170 NK=N9+K9+1-NR
2180 IFNK<=0THEN2200
2190 A(NR,NK)=A(NR,NK)+SK(II,LL)
2200 NEXTK9
2210 NEXTKK
2220 NEXTJ9,JJ
2230 NEXTME
2240 FORII=1TONF
2250 A(NS(II),1)=A(NS(II),1)*1E12+1E12
2255 Q(NS(II))=0
2260 NEXTII
2270 FORII=1TON6
2280 C(II)=Q(II)
2290 NEXTII
2300 PRINT:PRINT"THE SIMULTANEOUS EQUATIONS ARE NOW BEING SOLVED":PRINT
2310 GOSUB3710
2320 PRINT:PRINT"THE NODAL DISPLACEMENTS IN GLOBAL COORDS ARE AS FOLLOWS:":PRIN
T
2330 FORII=1TON6
```

```
2340 U(II)=C(II)
2350 PRINTU(II);
2360 NEXTII
2370 GOTO2950
2380 XA(1)=X(I)
2390 XA(2)=X(J)
2400 XA(3)=X(K)
2410 YA(1)=Y(I)
2420 YA(2)=Y(J)
2430 YA(3)=Y(K)
2440 ZA(1)=Z(I)
2450 ZA(2)=Z(J)
2460 ZA(3)=Z(K)
2470 S2(1)=XA(2)-XA(1)
2480 S2(2)=YA(2)-YA(1)
2490 S2(3)=ZA(2)-ZA(1)
2500 S3(1)=XA(3)-XA(1)
2510 S3(2)=YA(3)-YA(1)
2520 S3(3)=ZA(3)-ZA(1)
2530 L2=SQR(S2(1)↑2+S2(2)↑2+S2(3)↑2)
2540 Z1(1)=S2(1)/L2
2550 Z1(2)=S2(2)/L2
2560 Z1(3)=S2(3)/L2
2570 FORII=1TO3
2580 FORJJ=1TO3
2590 ZP(II,JJ)=0
2600 FORKK=1TO1
2610 ZP(II,JJ)=ZP(II,JJ)+Z1(II)*Z1(JJ)
2620 NEXTKK
2630 NEXTJJ,II
2640 FORII=1TO3
2650 FORJJ=1TO3
2660 UN=0
2670 IFII=JJTHENUN=1
2680 ZP(II,JJ)=UN-ZP(II,JJ)
2690 NEXTJJ,II
2700 FORII=1TO3
2710 S4(II)=0
2720 FORJJ=1TO3
2730 S4(II)=S4(II)+ZP(II,JJ)*S3(JJ)
2740 NEXTJJ,II
2750 L2=SQR(S4(1)↑2+S4(2)↑2+S4(3)↑2)
2760 Z2(1)=S4(1)/L2
2770 Z2(2)=S4(2)/L2
2780 Z2(3)=S4(3)/L2
2790 XY=(XA(1)*(YA(2)-YA(3))-YA(1)*(XA(2)-XA(3))+(XA(2)*YA(3)-YA(2)*XA(3)))/2
2800 YZ=(YA(1)*(ZA(2)-ZA(3))-ZA(1)*(YA(2)-YA(3))+(YA(2)*ZA(3)-ZA(2)*YA(3)))/2
2810 ZX=(ZA(1)*(XA(2)-XA(3))-XA(1)*(ZA(2)-ZA(3))+(ZA(2)*XA(3)-XA(2)*ZA(3)))/2
2820 AR=SQR(YZ↑2+ZX↑2+XY↑2)
2830 Z3(1)=YZ/AR
2840 Z3(2)=ZX/AR
2850 Z3(3)=XY/AR
2860 FORII=1TO3
2870 DC(1,II)=Z1(II)
2880 DC(2,II)=Z2(II)
2890 DC(3,II)=Z3(II)
2900 NEXTII
2910 FORII=1TO3
2920 FORJJ=1TO3:DT(JJ,II)=DC(II,JJ)
2930 NEXTJJ,II
2940 RETURN
2950 PRINT:PRINT"CALCULATION OF STRESSES HAS COMMENCED":PRINT
2960 FORME=1TOMS
2970 M8=3*ME
2980 I=IJ(M8-2)
2990 J=IJ(M8-1)
3000 K=IJ(M8)
3010 IFML>1THEN3090
3020 E=E(1)
3030 G=G(1)
3040 A=AR(1)
3050 IX=IX(1)
3060 IY=IY(1)
3070 IZ=IZ(1)
3080 GOTO3150
```

```
3090 E=E(NM(ME))
3100 G=G(NM(ME))
3110 A=AR(NM(ME))
3120 IX=IX(NM(ME))
3130 IY=IY(NM(ME))
3140 IZ=IZ(NM(ME))
3150 LE=SQR((X(J)-X(I))↑2+(Y(J)-Y(I))↑2+(Z(J)-Z(I))↑2)
3160 GOSUB2380
3170 I3=6*I-6
3180 J3=6*J-6
3190 FORII=1TO6
3200 UG(II)=U(I3+II)
3210 UG(II+6)=U(J3+II)
3220 NEXTII
3230 FORLL=1TO4
3240 FORII=1TO3
3250 I9=LL*3-3+II
3260 UL(I9)=0
3270 FORJJ=1TO3
3280 J9=3*LL-3+JJ
3290 UL(I9)=UL(I9)+DC(II,JJ)*UG(J9)
3300 NEXTJJ
3310 NEXTII
3320 NEXTLL
3330 FORCC=1TO2
3340 XI=1
3350 IFCC=1THENXI=0
3360 BM(1,1)=-1
3370 BM(1,7)=1
3380 BM(2,4)=-1
3390 BM(2,10)=1
3400 BM(3,3)=(6-12*XI)/LE↑2
3410 BM(3,5)=(-4+6*XI)/LE
3420 BM(3,9)=-(6-12*XI)/LE↑2
3430 BM(3,11)=(-2+6*XI)/LE
3440 BM(4,2)=-BM(3,3)
3450 BM(4,6)=BM(3,5)
3460 BM(4,8)=-BM(3,9)
3470 BM(4,12)=BM(3,11)
3480 FORJJ=1TO12
3490 BM(1,JJ)=BM(1,JJ)*A*E/LE
3500 BM(2,JJ)=BM(2,JJ)*G*IX/LE
3510 BM(3,JJ)=BM(3,JJ)*E*IY
3520 BM(4,JJ)=BM(4,JJ)*E*IZ
3530 NEXTJJ
3540 PRINT:PRINT"TO CONTINUE PRESS ANY KEY":PRINT
3550 GETA$:IFA$=""THEN3550
3560 PRINT"THE FOLLOWING FOUR VALUES ARE 1)AXIAL FORCE 2)TORQUE 3)MOMENT ABO";
3570 PRINT"UT XY PLANE 4)MOMENT ABOUT XZ PLANE; FOR MEMBER ";I;"-";J;"-";K
3580 IFCC=1THENPRINT"THIS SET IS AT NODE ";I
3590 IFCC=2THENPRINT"THIS SET IS AT NODE ";J
3600 FORII=1TO4
3610 SR(II)=0
3620 FORJJ=1TO12
3630 SR(II)=SR(II)+BM(II,JJ)*UL(JJ)
3640 NEXTJJ,II
3650 FORII=1TO4
3660 PRINTSR(II);
3670 NEXTII
3680 NEXTCC
3690 NEXTME
3700 END
3710 FORII=1TON6
3720 IK=II
3730 FORJJ=2TONW
3740 IK=IK+1
3750 KN=A(II,JJ)/A(II,1)
3760 JK=0
3770 FORKK=JJTONW
3780 JK=JK+1
3790 A(IK,JK)=A(IK,JK)-KN*A(II,KK)
3800 NEXTKK
3810 A(II,JJ)=KN
3820 C(IK)=C(IK)-KN*C(II)
3830 NEXTJJ
```

```
3840 C(II)=C(II)/A(II,1)
3850 NEXTII
3860 FORZZ=2TON6
3870 II=N6-ZZ+1
3880 FORKK=2TONW
3890 JJ=II+KK-1
3900 C(II)=C(II)-A(II,KK)*C(JJ)
3910 NEXTKK,ZZ
3920 RETURN
READY.
```

APPENDIX 10

```
140 PRINT:PRINT"STATIC ANALYSIS OF GRILLAGES":PRINT
150 INPUT"FEED NO OF JOINTS";NJ
160 INPUT"FEED NO OF SUPPRESSED DISPLACEMENTS";NF
170 N3=NJ*3
180 INPUT"NO OF CONCENTRATED LOADS";NC
190 DIMDC(6,6),SK(6,6),SX(6,6),SP(6),SQ(6),SG(6),SL(6)
200 DIMBX(2,6),ST(2),X(NJ),Y(NJ),NS(NF),IJ(250)
210 PRINT"FEED NODAL COORDINATES":PRINT
220 FORI=1TONJ
230 PRINT"X COORD FOR NODE";I
240 INPUTX(I)
250 PRINT"Y COORD FOR NODE";I
260 INPUTY(I)
270 NEXTI
280 INPUT"ELASTIC MODULUS";E
290 INPUT"RIGIDITY MODULUS";G
300 INPUT"2ND MOMENT OF AREA FOR X DIRECTION MEMBERS";AX:PRINT
310 INPUT"2ND MOMENT OF AREA FOR Y DIRECTION MEMBERS";AY:PRINT
320 INPUT"TORSIONAL CONSTANT FOR X DIRECTION MEMBERS";TX:PRINT
330 INPUT"TORSIONAL CONSTANT FOR Y DIRECTION MEMBERS";TY:PRINT
340 INPUT"NO OF MEMBERS IN X DIRECTION";MX
350 INPUT"NO OF MEMBERS IN Y DIRECTION";MY
360 MS=MX+MY
370 INPUT"LOAD/UNIT LENGTH ON ANY X DIRECTION MEMBER";UX
380 INPUT"LOAD/UNIT LENGTH ON ANY Y DIRECTION MEMBER";UY
390 NW=0
400 FORME=1TOMS
410 IF ME>MX THEN 470
420 PRINT"FEED I NODE OF (X) MEMBER";ME
430 INPUTI
440 PRINT"FEED J NODE OF (X) MEMBER";ME
450 INPUTJ
460 GOTO510
470 PRINT"FEED I NODE OF (Y) MEMBER";ME
480 INPUTI
490 PRINT"FEED J NODE OF (Y) MEMBER";ME
500 INPUTJ
510 IFABS(I-J)>NWTHENNW=ABS(I-J)
520 M8=ME*2-1
530 IJ(M8)=I
540 IJ(M8+1)=J
550 NEXTME
560 NW=(NW+1)*3
570 NT=N3+NW
580 DIMA(NT,NW),Q(NT),C(NT),U(N3)
590 IFNC=0THEN680
600 PRINT:PRINT"FEED NODAL POSITIONS OF CONCENTRATED LOADS":PRINT
610 FORI=1TONC
620 PRINT:PRINT"NODAL POSITION FOR LOAD";I:INPUTPO
630 PO=3*PO-2
640 PRINT:PRINT"FEED VALUE OF LOAD";I
650 INPUTQC
660 Q(PO)=Q(PO)+QC
670 NEXTI
680 PRINT:PRINT"FEED DISPLACEMENT POSITIONS OF SUPPRESSED DISPLACEMENTS":PRINT
690 FORII=1TONF
700 PRINT:PRINT"FEED POSITION";II
710 INPUTNS(II)
720 NEXTII
730 FORME=1TOMS
740 PRINT:PRINT"MEMBER";ME;"UNDER COMPUTATION":PRINT
750 M8=ME*2-1
760 I=IJ(M8)
770 J=IJ(M8+1)
780 IFME>MXTHEN830
790 UD=UX
800 SA=AX
810 TC=TX
820 GOTO860
830 UD=UY
840 SA=AY
850 TC=TY
```

```
860 FORII=1TO6
870 FORJJ=1TO6
880 DC(II,JJ)=0
890 SK(II,JJ)=0:SX(II,JJ)=0
900 NEXTJJ,II
910 LE=SQR((X(J)-X(I))↑2+(Y(J)-Y(I))↑2)
920 DC(1,1)=1:DC(4,4)=1
930 DC(2,3)=(Y(J)-Y(I))/LE
940 DC(5,6)=DC(2,3)
950 DC(3,2)=-DC(2,3)
960 DC(6,5)=DC(3,2)
970 DC(2,2)=(X(J)-X(I))/LE
980 DC(3,3)=DC(2,2)
990 DC(5,5)=DC(2,2)
1000 DC(6,6)=DC(2,2)
1010 SK(1,1)=12*E*SA/LE↑3
1020 SK(4,4)=SK(1,1)
1030 SK(4,1)=-SK(1,1)
1040 SK(1,4)=-SK(1,1)
1050 SK(2,2)=G*TC/LE
1060 SK(5,5)=SK(2,2)
1070 SK(5,2)=-SK(2,2)
1080 SK(2,5)=-SK(2,2)
1090 SK(3,3)=4*E*SA/LE
1100 SK(6,6)=SK(3,3)
1110 SK(6,3)=SK(3,3)/2
1120 SK(3,6)=SK(6,3)
1130 SK(4,3)=6*E*SA/LE↑2
1140 SK(3,4)=SK(4,3)
1150 SK(6,4)=SK(4,3)
1160 SK(4,6)=SK(4,3)
1170 SK(3,1)=-SK(4,3)
1180 SK(1,3)=SK(3,1)
1190 SK(6,1)=SK(3,1)
1200 SK(1,6)=SK(3,1)
1210 FORII=1TO6
1220 FORJJ=1TO6
1230 SX(II,JJ)=0
1240 NEXTJJ
1250 FORJJ=1TO6
1260 FORKK=1TO6
1270 SX(II,JJ)=SX(II,JJ)+SK(II,KK)*DC(KK,JJ)
1280 NEXTKK,JJ,II
1290 FORII=1TO6
1300 FORJJ=1TO6
1310 SK(II,JJ)=0
1320 NEXTJJ
1330 FORJJ=1TO6
1340 FORKK=1TO6
1350 SK(II,JJ)=SK(II,JJ)+DC(KK,II)*SX(KK,JJ)
1360 NEXTKK,JJ,II
1370 SP(1)=UD*LE/2
1380 SP(4)=SP(1)
1390 SP(2)=0:SP(5)=0
1400 SP(3)=-UD*LE↑2/12
1410 SP(6)=-SP(3)
1420 FORII=1TO6
1430 SQ(II)=0
1440 FORJJ=1TO6
1450 SQ(II)=SQ(II)+DC(JJ,II)*SP(JJ)
1460 NEXTJJ,II
1470 I1=3*I-3
1480 J1=3*J-3
1490 Q(I1+1)=Q(I1+1)+SQ(1)
1500 Q(I1+2)=Q(I1+2)+SQ(2)
1510 Q(I1+3)=Q(I1+3)+SQ(3)
1520 Q(J1+1)=Q(J1+1)+SQ(4)
1530 Q(J1+2)=Q(J1+2)+SQ(5)
1540 Q(J1+3)=Q(J1+3)+SQ(6)
1550 FORJJ=1TO2
1560 NR=J1
1570 IFJJ=1THENNR=I1
1580 FORJ9=1TO3
1590 NR=NR+1:II=(JJ-1)*3+J9
1600 FORKK=1TO2
```

```
1610 N9=J1
1620 IFKK=1THENN9=I1
1630 FORK=1TO3
1640 LL=(KK-1)*3+K:NK=N9+K+1-NR
1650 IFNK<=0THEN1670
1660 A(NR,NK)=A(NR,NK)+SK(II,LL)
1670 NEXTK
1680 NEXTKK,J9,JJ
1690 NEXTME
1700 FORI=1TONF
1710 A(NS(I),1)=A(NS(I),1)*1E12+1E12
1715 Q(NS(I))=0
1720 NEXTI
1730 FORII=1TON3
1740 C(II)=Q(II)
1750 NEXTII
1760 N=N3
1770 PRINT:PRINT"THE SIMULTANEOUS EQUATIONS ARE NOW BEING SOLVED":PRINT
1780 GOSUB2470
1790 FORI=1TON3
1800 U(I)=C(I)
1810 NEXTI
1820 PRINT:PRINT"THE NODAL DISPLACEMENTS ARE AS FOLLOWS":PRINT
1830 FORI=1TON3
1840 PRINTU(I);
1850 NEXTI
1860 FORME=1TOMS
1870 M8=ME*2-1
1880 I=IJ(M8)
1890 J=IJ(M8+1)
1900 IFME>MXTHEN1950
1910 UD=UX
1920 SA=AX
1930 TC=TX
1940 GOTO1980
1950 UD=UY
1960 SA=AY
1970 TC=TY
1980 LE=SQR((X(J)-X(I))↑2+(Y(J)-Y(I))↑2)
1990 I1=3*I-3
2000 J1=3*J-3
2010 FORII=1TO3
2020 MM=I1+II
2030 NM=J1+II
2040 SG(II)=U(MM)
2050 SG(II+3)=U(NM)
2060 NEXTII
2070 FORII=1TO6
2080 FORJJ=1TO6
2090 DC(II,JJ)=0
2100 NEXTJJ,II
2110 DC(1,1)=1:DC(4,4)=1
2120 DC(2,3)=(Y(J)-Y(I))/LE
2130 DC(5,6)=DC(2,3)
2140 DC(3,2)=-DC(2,3)
2150 DC(6,5)=DC(3,2)
2160 DC(2,2)=(X(J)-X(I))/LE
2170 DC(3,3)=DC(2,2)
2180 DC(5,5)=DC(2,2)
2190 DC(6,6)=DC(2,2)
2200 FORII=1TO6
2210 SL(II)=0
2220 FORJJ=1TO6
2230 SL(II)=SL(II)+DC(II,JJ)*SG(JJ)
2240 NEXTJJ,II
2250 FORII=1TO2
2260 XI=1
2270 IFII=1THENXI=0
2280 BX(1,2)=-G*TC/LE
2290 BX(1,5)=-BX(1,2)
2300 BX(2,1)=E*SA*(6-12*XI)/LE↑2
2310 BX(2,3)=E*SA*(-4+6*XI)/LE
2320 BX(2,6)=E*SA*(-2+6*XI)/LE
2330 BX(2,4)=-BX(2,1)
2340 FORJJ=1TO2
```

```
2350 ST(JJ)=0
2360 FORKK=1TO6
2370 ST(JJ)=ST(JJ)+BX(JJ,KK)*SL(KK)
2380 NEXTKK,JJ
2390 ST(2)=ST(2)-UD*LE↑2/12
2400 PRINT:PRINT"TORQUE & BENDING MOMENT FOR MEMBER";I;"-";J;"AT XI=";XI
2410 PRINT"TORQUE=";ST(1),"MOMENT=";ST(2):PRINT
2420 PRINT"TO CONTINUE, PRESS ANY KEY":PRINT
2430 GETA$:IFA$=""THEN2430
2440 NEXTII
2450 NEXTME
2460 END
2470 FORII=1TON
2480 IK=II
2490 FORJJ=2TONW
2500 IK=IK+1
2510 CN=A(II,JJ)/A(II,1)
2520 JK=0
2530 FORKK=JJTONW
2540 JK=JK+1
2550 A(IK,JK)=A(IK,JK)-CN*A(II,KK)
2560 NEXTKK
2570 A(II,JJ)=CN
2580 C(IK)=C(IK)-CN*C(II)
2590 NEXTJJ
2600 C(II)=C(II)/A(II,1)
2610 NEXTII
2620 FORIZ=2TON
2630 II=N-IZ+1
2640 FORKK=2TONW
2650 JJ=II+KK-1
2660 C(II)=C(II)-A(II,KK)*C(JJ)
2670 NEXTKK,IZ
2680 RETURN
READY.
```

APPENDIX 11

```
140 PRINT:PRINT"STATIC ANALYSIS OF GRILLAGES UNDER HYDROSTATIC LOADING":PRINT
150 INPUT"FEED NO OF ELEMENTS";MS
160 INPUT"FEED NO OF JOINTS";NJ
170 INPUT"FEED NO OF SUPPRESSED DISPLACEMENTS";NF
180 N3=NJ*3
190 INPUT"NO OF CONCENTRATED LOADS";NC
200 DIMDC(6,6),SK(6,6),SX(6,6),SP(6),SQ(6),SG(6),SL(6)
210 DIMBX(2,6),ST(2),X(NJ),Y(NJ),NS(NF),IJ(250)
220 DIMSA(MS),TC(MS),U0(MS),UL(MS)
230 PRINT"FEED NODAL COORDINATES":PRINT
240 FORI=1TONJ
250 PRINT"X COORD FOR NODE";I
260 INPUTX(I)
270 PRINT"Y COORD FOR NODE";I
280 INPUTY(I)
290 NEXTI
300 INPUT"ELASTIC MODULUS";E
310 INPUT"RIGIDITY MODULUS";G
320 NW=0
330 FORME=1TOMS
340 PRINT"FEED I NODE OF MEMBER";ME
350 INPUTI
360 PRINT"FEED J NODE OF MEMBER";ME
370 INPUTJ
380 IFABS(I-J)>NWTHENNW=ABS(I-J)
390 M8=ME*2-1
400 IJ(M8)=I
410 IJ(M8+1)=J
420 PRINT:PRINT"FEED 2ND MOMENT OF AREA FOR MEMBER";I;"-";J
430 INPUTSA(ME)
440 PRINT:PRINT"FEED TORSIONAL CONSTANT FOR MEMBER";I;"-";J
450 INPUTTC(ME)
460 PRINT:PRINT"FEED LOAD/UNIT LENGTH AT NODE";I
470 INPUTU0(ME)
480 PRINT:PRINT"FEED LOAD/UNIT LENGTH AT NODE";J
490 INPUTUL(ME)
500 NEXTME
510 NW=(NW+1)*3
520 NT=N3+NW
530 DIMA(NT,NW),Q(NT),C(NT),U(N3)
540 IFNC=0THEN630
550 PRINT:PRINT"FEED NODAL POSITIONS OF CONCENTRATED LOADS":PRINT
560 FORI=1TONC
570 PRINT:PRINT"NODAL POSITION FOR LOAD";I:INPUTPO
580 PO=3*PO-2
590 PRINT:PRINT"FEED VALUE OF LOAD";I
600 INPUTQC
610 Q(PO)=Q(PO)+QC
620 NEXTI
630 PRINT:PRINT"FEED DISPLACEMENT POSITIONS OF SUPPRESSED DISPLACEMENTS":PRINT
640 FORII=1TONF
650 PRINT:PRINT"FEED POSITION";II
660 INPUTNS(II)
670 NEXTII
680 FORME=1TOMS
690 PRINT:PRINT"MEMBER";ME;"UNDER COMPUTATION":PRINT
700 M8=ME*2-1
710 I=IJ(M8)
720 J=IJ(M8+1)
730 SA=SA(ME)
740 TC=TC(ME)
750 U0=U0(ME)
760 UL=UL(ME)
770 FORII=1TO6
780 FORJJ=1TO6
790 DC(II,JJ)=0
800 SK(II,JJ)=0:SX(II,JJ)=0
810 NEXTJJ,II
820 LE=SQR((X(J)-X(I))↑2+(Y(J)-Y(I))↑2)
830 DC(1,1)=1:DC(4,4)=1
840 DC(2,3)=(Y(J)-Y(I))/LE
850 DC(5,6)=DC(2,3)
```

```
860 DC(3,2)=-DC(2,3)
870 DC(6,5)=DC(3,2)
880 DC(2,2)=(X(J)-X(I))/LE
890 DC(3,3)=DC(2,2)
900 DC(5,5)=DC(2,2)
910 DC(6,6)=DC(2,2)
920 SK(1,1)=12*E*SA/LE↑3
930 SK(4,4)=SK(1,1)
940 SK(4,1)=-SK(1,1)
950 SK(1,4)=-SK(1,1)
960 SK(2,2)=G*TC/LE
970 SK(5,5)=SK(2,2)
980 SK(5,2)=-SK(2,2)
990 SK(2,5)=-SK(2,2)
1000 SK(3,3)=4*E*SA/LE
1010 SK(6,6)=SK(3,3)
1020 SK(6,3)=SK(3,3)/2
1030 SK(3,6)=SK(6,3)
1040 SK(4,3)=6*E*SA/LE↑2
1050 SK(3,4)=SK(4,3)
1060 SK(6,4)=SK(4,3)
1070 SK(4,6)=SK(4,3)
1080 SK(3,1)=-SK(4,3)
1090 SK(1,3)=SK(3,1)
1100 SK(6,1)=SK(3,1)
1110 SK(1,6)=SK(3,1)
1120 FORII=1TO6
1130 FORJJ=1TO6
1140 SX(II,JJ)=0
1150 NEXTJJ
1160 FORJJ=1TO6
1170 FORKK=1TO6
1180 SX(II,JJ)=SX(II,JJ)+SK(II,KK)*DC(KK,JJ)
1190 NEXTKK,JJ,II
1200 FORII=1TO6
1210 FORJJ=1TO6
1220 SK(II,JJ)=0
1230 NEXTJJ
1240 FORJJ=1TO6
1250 FORKK=1TO6
1260 SK(II,JJ)=SK(II,JJ)+DC(KK,II)*SX(KK,JJ)
1270 NEXTKK,JJ,II
1280 SP(1)=7*U0*LE/20+3*UL*LE/20
1290 SP(4)=3*U0*LE/20+7*UL*LE/20
1300 SP(2)=0:SP(5)=0
1310 SP(3)=-(U0/20+UL/30)*LE*LE
1320 SP(6)=(U0/30+UL/20)*LE*LE
1330 FORII=1TO6
1340 SQ(II)=0
1350 FORJJ=1TO6
1360 SQ(II)=SQ(II)+DC(JJ,II)*SP(JJ)
1370 NEXTJJ,II
1380 I1=3*I-3
1390 J1=3*J-3
1400 Q(I1+1)=Q(I1+1)+SQ(1)
1410 Q(I1+2)=Q(I1+2)+SQ(2)
1420 Q(I1+3)=Q(I1+3)+SQ(3)
1430 Q(J1+1)=Q(J1+1)+SQ(4)
1440 Q(J1+2)=Q(J1+2)+SQ(5)
1450 Q(J1+3)=Q(J1+3)+SQ(6)
1460 FORJJ=1TO2
1470 NR=J1
1480 IFJJ=1THENNR=I1
1490 FORJ9=1TO3
1500 NR=NR+1:II=(JJ-1)*3+J9
1510 FORKK=1TO2
1520 N9=J1
1530 IFKK=1THENN9=I1
1540 FORK=1TO3
1550 LL=(KK-1)*3+K:NK=N9+K+1-NR
1560 IFNK<=0THEN1580
1570 A(NR,NK)=A(NR,NK)+SK(II,LL)
1580 NEXTK
1590 NEXTKK,J9,JJ
1600 NEXTME
```

```
1610 FORI=1TONF
1620 A(NS(I),1)=A(NS(I),1)*1E12+1E12
1625 Q(NS(I))=0
1630 NEXTI
1640 FORII=1TON3
1650 C(II)=Q(II)
1660 NEXTII
1670 N=N3
1680 PRINT:PRINT"THE SIMULTANEOUS EQUATIONS ARE NOW BEING SOLVED":PRINT
1690 GOSUB2350
1700 FORI=1TON3
1710 U(I)=C(I)
1720 NEXTI
1730 PRINT:PRINT"THE NODAL DISPLACEMENTS ARE AS FOLLOWS":PRINT
1740 FORI=1TON3
1750 PRINTU(I);
1760 NEXTI
1770 FORME=1TOMS
1780 M8=ME*2-1
1790 I=IJ(M8)
1800 J=IJ(M8+1)
1810 SA=SA(ME)
1820 TC=TC(ME)
1830 U0=U0(ME)
1840 UL=UL(ME)
1850 LE=SQR((X(J)-X(I))↑2+(Y(J)-Y(I))↑2)
1860 I1=3*I-3
1870 J1=3*J-3
1880 FORII=1TO3
1890 MM=I1+II
1900 NM=J1+II
1910 SG(II)=U(MM)
1920 SG(II+3)=U(NM)
1930 NEXTII
1940 FORII=1TO6
1950 FORJJ=1TO6
1960 DC(II,JJ)=0
1970 NEXTJJ,II
1980 DC(1,1)=1:DC(4,4)=1
1990 DC(2,3)=(Y(J)-Y(I))/LE
2000 DC(5,6)=DC(2,3)
2010 DC(3,2)=-DC(2,3)
2020 DC(6,5)=DC(3,2)
2030 DC(2,2)=(X(J)-X(I))/LE
2040 DC(3,3)=DC(2,2)
2050 DC(5,5)=DC(2,2)
2060 DC(6,6)=DC(2,2)
2070 FORII=1TO6
2080 SL(II)=0
2090 FORJJ=1TO6
2100 SL(II)=SL(II)+DC(II,JJ)*SG(JJ)
2110 NEXTJJ,II
2120 FORII=1TO2
2130 XI=1
2140 IFII=1THENXI=0
2150 BX(1,2)=-G*TC/LE
2160 BX(1,5)=-BX(1,2)
2170 BX(2,1)=E*SA*(6-12*XI)/LE↑2
2180 BX(2,3)=E*SA*(-4+6*XI)/LE
2190 BX(2,6)=E*SA*(-2+6*XI)/LE
2200 BX(2,4)=-BX(2,1)
2210 FORJJ=1TO2
2220 ST(JJ)=0
2230 FORKK=1TO6
2240 ST(JJ)=ST(JJ)+BX(JJ,KK)*SL(KK)
2250 NEXTKK,JJ
2260 IFXI=0THENST(2)=ST(2)-(U0/20+UL/30)*LE*LE
2270 IFXI=1THENST(2)=ST(2)-(U0/30+UL/20)*LE*LE
2280 PRINT:PRINT"TORQUE & BENDING MOMENT FOR MEMBER";I;"-";J;"AT XI=";XI
2290 PRINT"TORQUE=";ST(1),"MOMENT=";ST(2):PRINT
2300 PRINT"TO CONTINUE, PRESS ANY KEY":PRINT
2310 GETA$:IFA$=""THEN2310
2320 NEXTII
2330 NEXTME
2340 END
```

```
 2350 FORII=1TON
 2360 IK=II
 2370 FORJJ=2TONW
 2380 IK=IK+1
 2390 CN=A(II,JJ)/A(II,1)
 2400 JK=0
 2410 FORKK=JJTONW
 2420 JK=JK+1
 2430 A(IK,JK)=A(IK,JK)-CN*A(II,KK)
 2440 NEXTKK
 2450 A(II,JJ)=CN
 2460 C(IK)=C(IK)-CN*C(II)
 2470 NEXTJJ
 2480 C(II)=C(II)/A(II,1)
 2490 NEXTII
 2500 FORIZ=2TON
 2510 II=N-IZ+1
 2520 FORKK=2TONW
 2530 JJ=II+KK-1
 2540 C(II)=C(II)-A(II,KK)*C(JJ)
 2550 NEXTKK,IZ
 2560 RETURN
READY.
```

APPENDIX 12

```
 10 PRINT:PRINT:PRINT"AXISYMMETRIC STRESSES IN THIN WALLED TAPERED CONICAL SHELL
S";
 15 PRINT" UNDER VARYING LATERAL PRESSURE AND AN AXIAL PRESSURE":PRINT:PRINT
 20 PRINT "FEED IN NUMBER OF ELEMENTS"
 30 INPUT ES
 40 PRINT "NUMBER OF SUPPRESSED DISPLTS"
 50 INPUT NF
 60 N=3*ES+3
 70 NN=ES+1
 80 NW=6
 90 NT=N+NW
 100 DIM K(6,6),DC(6,6),B(4,6)
 110 DIM SG(6),SL(6),SS(4),MX(6),PP(6)
 120 DIM XL(NN),RI(NN),TT(NN),PR(NN)
 130 DIM BT(6,4),D(4,4),DB(4,6),BB(6,6)
 140 DIM A(NT,NW),C(NT),Q(N),DI(N),NS(NF)
 150 PRINT "ELASTIC MODULUS"
 160 INPUT E
 170 PRINT "POISSONS RATIO"
 180 INPUT NU
 190 PRINT "END PRESSURE"
 200 INPUT PX
 210 PRINT "POSNS OF SUPPRESSED DISPLTS"
 220 FOR II=1 TO NF
 230 PRINT "POSITION";II
 240 INPUT NS(II)
 250 NEXT II
 260 FOR II=1 TO NN
 270 PRINT "X COORD FOR NODE";II
 280 INPUT XL(II)
 290 PRINT "Y COORD FOR NODE";II
 300 INPUT RI(II)
 310 PRINT "THICKNESS AT NODE";II
 320 INPUT TT(II)
 330 PRINT "LATERAL PRESSURE AT NODE";II
 340 INPUT PR(II)
 350 NEXT II
 360 FOR EL=1 TO ES
 365 PRINT:PRINT:PRINT"ELEMENT ";EL;" UNDER COMPUTATION"
 370 I=EL
 380 J=EL+1
 390 FOR II=1 TO 6
 400 SL(II)=0
 410 FOR JJ=1 TO 6
 420 K(II,JJ)=0
 430 DC(II,JJ)=0
 440 NEXT JJ
 450 NEXT II
 460 R1=RI(I)
 470 R2=RI(I+1)
 480 LL=XL(I+1)-XL(I)
 490 L=SQR((R2-R1)↑2+LL↑2)
 500 TH=ATN((R2-R1)/LL)
 510 SY=SIN(TH)
 520 KS=COS(TH)
 530 FOR IJ=1 TO 4
 535 LW=0
 540 UP=1
 545 GOSUB 6500
 550 R=R1+R2*XI-R1*XI
 560 PR=PR(I)*(1-XI)+PR(I+1)*XI
 570 T=TT(I)*(1-XI)+TT(I+1)*XI
 580 GOSUB 7000
 585 GOSUB 7510
 590 FOR II=1 TO 4
 600 FOR JJ=1 TO 6
 610 BT(JJ,II)=B(II,JJ)
 620 NEXT JJ
 630 NEXT II
 640 FOR II=1 TO 6
 650 FOR JJ=1 TO 6
 660 BB(II,JJ)=0
```

```
670 NEXT JJ
680 NEXT II
690 FOR II=1 TO 6
700 FOR JJ=1 TO 6
710 FOR KK=1 TO 4
720 BB(II,JJ)=BB(II,JJ)+BT(II,KK)*DB(KK,JJ)
730 NEXT KK
740 BB(II,JJ)=BB(II,JJ)*R*AI
750 NEXT JJ
760 NEXT II
770 FOR II=1 TO 6
780 SL(II)=SL(II)+MX(II)
790 FOR JJ=1 TO 6
800 K(II,JJ)=K(II,JJ)+BB(II,JJ)
810 NEXT JJ
820 NEXT II
830 NEXT IJ
835 CN=2*π*L
840 FOR II=1 TO 6
845 FOR JJ=1 TO 6
850 K(II,JJ)=K(II,JJ)*CN
855 NEXT JJ
860 SL(II)=SL(II)*CN
865 NEXT II
870 DC(1,1)=KS
880 DC(2,2)=KS
890 DC(4,4)=KS
900 DC(5,5)=KS
910 DC(1,2)=SY
920 DC(4,5)=SY
930 DC(2,1)=-SY
940 DC(5,4)=-SY
950 DC(3,3)=1
960 DC(6,6)=1
970 FOR II=1 TO 6
980 FOR JJ=1 TO 6
990 BB(II,JJ)=0
1000 NEXT JJ
1010 FOR JJ=1 TO 6
1020 FOR KK=1 TO 6
1030 BB(II,JJ)=BB(II,JJ)+K(II,KK)*DC(KK,JJ)
1040 NEXT KK
1050 NEXT JJ
1060 NEXT II
1070 FOR II=1 TO 6
1080 FOR JJ=1 TO 6
1090 K(II,JJ)=0
1100 NEXT JJ
1110 FOR JJ=1 TO 6
1120 FOR KK=1 TO 6
1130 K(II,JJ)=K(II,JJ)+DC(KK,II)*BB(KK,JJ)
1140 NEXT KK
1150 NEXT JJ
1160 NEXT II
1170 I1=3*I-3
1180 J1=3*J-3
1190 FOR II=1 TO 6
1200 PP(II)=0
1210 FOR JJ=1 TO 6
1220 PP(II)=PP(II)+DC(JJ,II)*SL(JJ)
1230 NEXT JJ
1240 NEXT II
1250 Q(I1+1)=Q(I1+1)+PP(1)
1260 Q(I1+2)=Q(I1+2)+PP(2)
1270 Q(I1+3)=Q(I1+3)+PP(3)
1280 Q(J1+1)=Q(J1+1)+PP(4)
1290 Q(J1+2)=Q(J1+2)+PP(5)
1300 Q(J1+3)=Q(J1+3)+PP(6)
1310 FOR II=1 TO 6
1320 FOR JJ=II TO 6
1330 MG=I1+II
1340 TR=I1+JJ-MG+1
1350 A(MG,TR)=A(MG,TR)+K(II,JJ)
1360 NEXT JJ
1370 NEXT II
```

```
1380 NEXT EL
1390 UN=1
1400 IF PX>0 THEN UN=-1
1410 Q(N-2)=Q(N-2)+PX*π*(RI(NN)+UN*TT(NN)/2)↑2
1420 FOR II=1 TO NF
1430 A(NS(II),1)=A(NS(II),1)*1E14+1E14
1440 Q(NS(II))=0
1450 NEXT II
1460 FOR II=1 TO N
1470 C(II)=Q(II)
1480 NEXT II
1490 GOSUB 6000
1500 PRINT:PRINT:PRINT"THE NODAL DISPLACEMENTS U V AND THETA ARE AS FOLLOWS":PR
INT
1510 FOR II=1 TO N
1520 DI(II)=C(II)
1530 PRINT DI(II);
1540 NEXT II
1545 PRINT
1550 FOR EL=1 TO ES
1560 I1=3*EL-3
1570 I=EL
1580 R1=RI(I)
1590 R2=RI(I+1)
1600 LL=XL(I+1)-XL(I)
1610 L=SQR((R2-R1)↑2+LL↑2)
1620 TH=ATN((R2-R1)/LL)
1630 SY=SIN(TH)
1640 KS=COS(TH)
1650 FOR II=1 TO 6
1660 FOR JJ=1 TO 6
1670 DC(II,JJ)=0
1680 NEXT JJ
1690 NEXT II
1700 DC(1,1)=KS
1710 DC(2,2)=KS
1720 DC(4,4)=KS
1730 DC(5,5)=KS
1740 DC(1,2)=SY
1750 DC(2,1)=-SY
1760 DC(4,5)=SY
1770 DC(5,4)=-SY
1780 DC(3,3)=1
1790 DC(6,6)=1
1800 FOR II=1 TO 6
1810 MM=I1+II
1820 SG(II)=DI(MM)
1830 NEXT II
1840 FOR II=1 TO 6
1850 SL(II)=0
1860 FOR JJ=1 TO 6
1870 SL(II)=SL(II)+DC(II,JJ)*SG(JJ)
1880 NEXT JJ
1890 NEXT II
1900 XI=-1
1910 FOR KK=1 TO 2
1920 XI=XI+1
1930 R=R1+R2*XI-R1*XI
1940 T=TT(I)*(1-XI)+TT(I+1)*XI
1950 IF R<T/100 THEN 2140
1960 GOSUB 7000
1970 FOR II=1 TO 4
1980 SS(II)=0
1990 FOR JJ=1 TO 6
2000 SS(II)=SS(II)+DB(II,JJ)*SL(JJ)
2010 NEXT JJ
2020 NEXT II
2030 SS(1)=SS(1)/T
2040 SS(2)=SS(2)/T
2050 CN=6/T↑2
2060 SS(3)=CN*SS(3)
2070 SS(4)=CN*SS(4)
2080 ET=SS(1)+SS(3)
2090 CM=SS(2)+SS(4)
2100 EM=SS(2)-SS(4)
```

```
2110 EN=SS(1)-SS(3)
2115 PRINT:PRINT"THE MERIDIONAL & HOOP STRESSES FOR THIS ELEMENT ARE";
2116 PRINT" GIVEN BELOW AT XI=";XI:PRINT
2120 PRINT "ELEM=";EL,"MERIDIN=";ET,"HOOPIN=";CM,"MERIDEX=";EN,"HOOPEX=";EM
2125 PRINT:PRINT:PRINT"TO CONTINUE PRESS ANY KEY":PRINT:PRINT
2130 GET A$:IF A$="" GOTO 2130
2140 NEXT KK
2150 NEXT EL
2160 END
6000 FOR II=1 TO N
6010 IK=II
6020 FOR JJ=2 TO NW
6030 IK=IK+1
6040 CN=A(II,JJ)/A(II,1)
6050 JK=0
6060 FOR KK=JJ TO NW
6070 JK=JK+1
6080 A(IK,JK)=A(IK,JK)-CN*A(II,KK)
6090 NEXT KK
6100 A(II,JJ)=CN
6110 C(IK)=C(IK)-CN*C(II)
6120 NEXT JJ
6130 C(II)=C(II)/A(II,1)
6140 NEXT II
6150 FORIZ=2TON
6160 II=N-IZ+1
6170 FORKK=2TONW
6180 JJ=II+KK-1
6190 C(II)=C(II)-A(II,KK)*C(JJ)
6200 NEXTKK,IZ
6210 RETURN
6500 Z9=0.5*(UP-LW)
6505 IF IJ=1 THEN 6520
6510 GOTO 6540
6520 XI=-0.8611361
6530 AI=0.34785485
6540 IF IJ=2 THEN 6560
6550 GOTO 6580
6560 XI=-0.33998104
6570 AI=0.65214515
6580 IF IJ=3 THEN 6600
6590 GOTO 6620
6600 XI=0.33998104
6610 AI=0.65214515
6620 IF IJ=4 THEN 6640
6630 GOTO 6660
6640 XI=0.86113631
6650 AI=0.34785485
6660 XI=(1+XI)*Z9+LW
6665 AI=AI*Z9
6670 RETURN
6680 END
7000 FOR II=1 TO 4
7010 FOR JJ=1 TO 6
7015 B(II,JJ)=0
7020 NEXT JJ
7025 NEXT II
7030 B(1,1)=-1/L
7040 B(1,4)=1/L
7050 B(2,1)=(1-XI)*SY/R
7060 B(2,2)=(1-3*XI↑2+2*XI↑3)*KS/R
7070 B(2,3)=L*(XI-2*XI↑2+XI↑3)*KS/R
7080 B(2,4)=XI*SY/R
7090 B(2,5)=(3*XI↑2-2*XI↑3)*KS/R
7100 B(2,6)=L*(-XI↑2+XI↑3)*KS/R
7110 B(3,2)=(-6+12*XI)/L↑2
7120 B(3,3)=(-4+6*XI)/L
7130 B(3,5)=(6-12*XI)/L↑2
7140 B(3,6)=(-2+6*XI)/L
7150 B(4,2)=(6*XI-6*XI↑2)*SY/(R*L)
7160 B(4,3)=(-1+4*XI-3*XI↑2)*SY/R
7170 B(4,5)=(-6*XI+6*XI↑2)*SY/(R*L)
7180 B(4,6)=(2*XI-3*XI↑2)*SY/R
7190 FOR II=1 TO 6
7200 B(4,II)=-B(4,II)
```

```
7210 NEXT II
7220 FOR II=1 TO 4
7230 FOR JJ=1 TO 4
7240 D(II,JJ)=0
7250 NEXT JJ
7260 NEXT II
7270 D(1,1)=1
7280 D(1,2)=NU
7290 D(2,1)=NU
7300 D(2,2)=1
7310 D(3,3)=T↑2/12
7320 D(4,4)=T↑2/12
7330 D(3,4)=NU*T↑2/12
7340 D(4,3)=NU*T↑2/12
7350 ET=E*T/(1-NU↑2)
7360 FOR II=1 TO 4
7370 FOR JJ=1 TO 4
7380 D(II,JJ)=D(II,JJ)*ET
7390 NEXT JJ
7400 NEXT II
7410 FOR II=1 TO 4
7420 FOR JJ=1 TO 6
7430 DB(II,JJ)=0
7440 NEXT JJ
7450 FOR JJ=1 TO 6
7460 FOR IK=1 TO 4
7470 DB(II,JJ)=DB(II,JJ)+D(II,IK)*B(IK,JJ)
7480 NEXT IK
7490 NEXT JJ
7495 NEXT II
7500 RETURN
7505 END
7510 MX(1)=0
7520 MX(4)=0
7530 MX(3)=L*(XI-2*XI↑2+XI↑3)
7540 MX(5)=3*XI↑2-2*XI↑3
7550 MX(2)=1-3*XI↑2+2*XI↑3
7560 MX(6)=L*(-XI↑2+XI↑3)
7570 CN=R*PR*AI
7580 FOR II=1 TO 6
7590 MX(II)=MX(II)*CN
7610 NEXT II
7620 RETURN
7630 END
READY.
```

APPENDIX 13

```
140 PRINT:PRINT"PROGRAM FOR PLANE STRESS & PLANE STRAIN":PRINT
150 PRINT"FEED NO OF ELEMENTS":PRINT
160 INPUTMS
170 M3=3*MS
180 PRINT"FEED NO OF NODES":PRINT
190 INPUTNN
200 N2=2*NN
210 PRINT"FEED NO OF SUPPRESSED DISPLACEMENTS":PRINT
220 INPUTNF
230 DIMSK(6,6),IJ(M3),NS(NF),UG(N2)
240 DIMB(3),C(3),U(3),V(3)
250 DIMBM(3,6),BD(3,6),D(3,3),SR(3)
260 PRINT"FEED ELASTIC MODULUS"
270 INPUTE
280 PRINT"FEED POISSONS RATIO"
290 INPUTNU
300 PRINT"FEED PLATE THICKNESS"
310 INPUTT
320 PRINT:PRINT"IF PLANE STRESS FEED 1,BUT IF PLANE STRAIN FEED ZERO":PRINT
330 INPUTSS
340 IFSS=0THEN390
350 CN=E*T/(1-NU↑2)
360 ZU=NU
370 MU=(1-NU)/2
380 GOTO420
390 CN=E*T*(1-NU)/((1+NU)*(1-2*NU))
400 ZU=NU/(1-NU)
410 MU=(1-2*NU)/(2*(1-NU))
420 PRINT:PRINT"FEED THE NODAL POINTS DESCRIBING EACH ELEMENT":PRINT
430 MX=0
440 FORME=1TOMS
450 PRINT"FEED I,J&K FOR ELEMENT";ME
460 INPUTIN,JN,KN
470 I8=3*ME
480 IJ(I8-2)=IN
490 IJ(I8-1)=JN
500 IJ(I8)=KN
510 IFABS(JN-IN)>MXTHENMX=ABS(JN-IN)
520 IFABS(KN-JN)>MXTHENMX=ABS(KN-JN)
530 IFABS(KN-IN)>MXTHENMX=ABS(KN-IN)
540 NEXTME
550 NW=(MX+1)*2
560 NT=NW+N2
570 DIMA(NT,NW),Q(NT),CV(NT),X(NN),Y(NN)
580 PRINT"FEED NODAL COORDINATES":PRINT
590 FORII=1TONN
600 PRINT"FEED X COORD FOR NODE";II
610 INPUTX(II)
620 PRINT"FEED Y COORD FOR NODE";II
630 INPUTY(II)
640 NEXTII
650 PRINT:PRINT"FEED THE NUMBER OF NODES WITH EXTERNAL LOADS":PRINT
660 INPUTNC
670 PRINT:PRINT"FEED THE POSNS AND VALUES OF EXTERNAL LOADS":PRINT
680 FORIJ=1TONC
690 PRINT"NODAL POSITION OF LOAD";IJ
700 INPUTNP
710 PRINT"VALUE OF HORIZONTAL COMPONENT"
720 INPUTWC
730 Q(2*NP-1)=WC
740 PRINT"VALUE OF VERTICAL COMPONENT"
750 INPUTWC
760 Q(2*NP)=WC
770 NEXTIJ
780 PRINT:PRINT"FEED (DISPLACEMENT) POSITIONS OF SUPPRESSED DISPLACEMENTS":PRIN
T
790 FORIJ=1TONF
800 PRINT"(POSITION)";IJ
810 INPUTN9
820 NS(IJ)=N9
830 NEXTIJ
840 FORME=1TOMS
```

```
845 PRINT"ELEMENT NO";ME;"UNDER COMPUTATION"
850 I8=3*ME
860 IN=IJ(I8-2)
870 JN=IJ(I8-1)
880 KN=IJ(I8)
890 B(1)=Y(JN)-Y(KN)
900 B(2)=Y(KN)-Y(IN)
910 B(3)=Y(IN)-Y(JN)
920 C(1)=X(KN)-X(JN)
930 C(2)=X(IN)-X(KN)
940 C(3)=X(JN)-X(IN)
950 DL=X(IN)*B(1)+X(JN)*B(2)+X(KN)*B(3)
960 DL=ABS(DL)
970 FORI=1TO3
980 I2=2*I-2
990 FORJ=1TO3
1000 J2=2*J-2
1010 SK(I2+1,J2+1)=0.5*CN*(B(I)*B(J)+MU*C(I)*C(J))/DL
1020 SK(I2+2,J2+2)=0.5*CN*(C(I)*C(J)+MU*B(I)*B(J))/DL
1030 SK(I2+1,J2+2)=0.5*CN*(ZU*B(I)*C(J)+MU*C(I)*B(J))/DL
1040 SK(I2+2,J2+1)=0.5*CN*(ZU*B(J)*C(I)+MU*C(J)*B(I))/DL
1050 NEXTJ,I
1060 I1=2*IN-2
1070 J1=2*JN-2
1080 K1=2*KN-2
1090 FORJJ=1TO3
1100 IFJJ=1THENNR=I1
1110 IFJJ=2THENNR=J1
1120 IFJJ=3THENNR=K1
1130 FORJ9=1TO2
1140 NR=NR+1:II=(JJ-1)*2+J9
1150 FORKK=1TO3
1160 IFKK=1THENN9=I1
1170 IFKK=2THENN9=J1
1180 IFKK=3THENN9=K1
1190 FORK=1TO2
1200 LL=(KK-1)*2+K
1210 NK=N9+K+1-NR
1220 IFNK<=0THEN1240
1230 A(NR,NK)=A(NR,NK)+SK(II,LL)
1240 NEXTK
1250 NEXTKK
1260 NEXTJ9
1270 NEXTJJ
1280 NEXTME
1290 FORII=1TONF
1300 N9=NS(II)
1310 A(N9,1)=A(N9,1)*1.0E12+1E12
1315 Q(N9)=0
1320 NEXTII
1330 FORII=1TON2
1340 CV(II)=Q(II)
1350 NEXTII
1355 PRINT"THE SIMULTANEOUS EQNS ARE NOW BEING SOLVED"
1360 GOSUB2060
1370 PRINT"THE NODAL DISPLACEMENTS IN GLOBAL COORDS ARE AS FOLLOWS:"
1380 FORI=1TON2
1390 UG(I)=CV(I)
1400 PRINTUG(I);
1410 NEXTI
1420 CN=CN/T
1430 PRINT:PRINT:PRINT"TO CONTINUE PRESS ANY KEY":PRINT
1440 GETA$:IFA$=""THEN1440
1450 FORME=1TOMS
1460 I8=3*ME
1470 IN=IJ(I8-2)
1480 JN=IJ(I8-1)
1490 KN=IJ(I8)
1500 U(1)=UG(2*IN-1)
1510 U(2)=UG(2*JN-1)
1520 U(3)=UG(2*KN-1)
1530 V(1)=UG(2*IN)
1540 V(2)=UG(2*JN)
1550 V(3)=UG(2*KN)
1560 B(1)=Y(JN)-Y(KN)
```

```
1570 B(2)=Y(KN)-Y(IN)
1580 B(3)=Y(IN)-Y(JN)
1590 C(1)=X(KN)-X(JN)
1600 C(2)=X(IN)-X(KN)
1610 C(3)=X(JN)-X(IN)
1620 DL=X(IN)*B(1)+X(JN)*B(2)+X(KN)*B(3)
1630 DL=ABS(DL)
1640 D(1,1)=CN
1650 D(2,2)=CN
1660 D(1,2)=ZU*CN
1670 D(2,1)=ZU*CN
1680 D(3,3)=MU*CN
1690 BM(1,1)=B(1)/DL
1700 BM(1,2)=B(2)/DL
1710 BM(1,3)=B(3)/DL
1720 BM(2,4)=C(1)/DL
1730 BM(2,5)=C(2)/DL
1740 BM(2,6)=C(3)/DL
1750 BM(3,1)=C(1)/DL
1760 BM(3,2)=C(2)/DL
1770 BM(3,3)=C(3)/DL
1780 BM(3,4)=B(1)/DL
1790 BM(3,5)=B(2)/DL
1800 BM(3,6)=B(3)/DL
1810 FORII=1TO3
1820 FORJJ=1TO6
1830 BD(II,JJ)=0
1840 FORKK=1TO3
1850 BD(II,JJ)=BD(II,JJ)+D(II,KK)*BM(KK,JJ)
1860 NEXTKK
1870 NEXTJJ,II
1880 FORII=1TO3
1890 SR(II)=0
1900 FORJJ=1TO6
1910 IFJJ>3THEN1940
1920 SR(II)=SR(II)+BD(II,JJ)*U(JJ)
1930 GOTO1950
1940 SR(II)=SR(II)+BD(II,JJ)*V(JJ-3)
1950 NEXTJJ
1960 NEXTII
1970 PRINT:PRINT"THE STRESSES IN ELEMENT";IN;"-";JN;"-";KN;"ARE AS FOLLOWS:"
1980 PRINT
1990 PRINT:PRINT"SIGMAX=";SR(1)
2000 PRINT"SIGMAY=";SR(2)
2010 PRINT"TAUXY=";SR(3)
2020 PRINT:PRINT"TO CONTINUE PRESS ANY KEY"
2030 GETA$:IFA$=""THEN2030
2040 NEXTME
2050 END
2060 FORII=1TON2
2070 IK=II
2080 FORJJ=2TONW
2090 IK=IK+1
2100 KN=A(II,JJ)/A(II,1)
2110 JK=0
2120 FORKK=JJTONW
2130 JK=JK+1
2140 A(IK,JK)=A(IK,JK)-KN*A(II,KK)
2150 NEXTKK
2160 A(II,JJ)=KN
2170 CV(IK)=CV(IK)-KN*CV(II)
2180 NEXTJJ
2190 CV(II)=CV(II)/A(II,1)
2200 NEXTII
2210 FORIZ=2TON2
2220 II=N2-IZ+1
2230 FORKK=2TONW
2240 JJ=II+KK-1
2250 CV(II)=CV(II)-A(II,KK)*CV(JJ)
2260 NEXTKK,IZ
2270 RETURN
READY.
```

APPENDIX 14

```
 140 PRINT:PRINT"PROGRAM FOR SOLVING 2D EQUATIONS OF THE LAPLACE & POISSON TYPE"
:PRINT
 150 PRINT"FEED NO OF ELEMENTS":PRINT
 160 INPUTMS
 170 M3=3*MS
 180 PRINT"FEED NO OF NODES":PRINT
 190 INPUTNN
 200 PRINT:PRINT"FEED THE NUMBER OF KNOWN VALUES OF THE FUNCTION ON THE BOUNDARY
 ";
 210 PRINT"NODES":PRINT
 220 INPUTNF
 230 DIMSK(3,3),IJ(M3),NS(NF),NQ(NF),UG(NN)
 240 DIMB(3),C(3),UL(3),BM(2,3),SL(2)
 250 PRINT:PRINT"IF THE VALUE OF (Q) IS CONSTANT OVER THE ENTIRE SURFACE, ";
 260 PRINT"FEED 1; ELSE FEED ZERO":PRINT
 270 INPUTQ7
 280 IFQ7=0THEN340
 290 IFQ7=1THEN320
 300 PRINT:PRINT"FEED 1 OR ZERO":PRINT
 310 GOTO250
 320 PRINT:PRINT"FEED Q":PRINT
 330 INPUTQ
 340 PRINT:PRINT"FEED THE VALUE OF THE CONSTANT K.":PRINT
 350 INPUTC8
 360 PRINT:PRINT"IF ANY BOUNDARY HAS A VALUE FOR HL, FEED 1, ELSE FEED ZERO":PRI
NT
 370 INPUTXH
 380 IFXH=1ORXH=0THEN410
 390 PRINT:PRINT"FEED 1 OR ZERO":PRINT
 400 GOTO360
 410 PRINT:PRINT"IF ANY BOUNDARY HAS A VALUE FOR QG, FEED 1, ELSE FEED ZERO":PRI
NT
 420 INPUTXQ
 430 IFXQ=1ORXQ=0THEN460
 440 PRINT:PRINT"FEED 1 OR ZERO":PRINT
 450 GOTO410
 460 PRINT:PRINT"FEED THE NODAL POINTS DESCRIBING EACH ELEMENT":PRINT
 470 MX=0
 480 FORME=1TOMS
 490 PRINT"FEED I,J&K FOR ELEMENT";ME
 500 INPUTIN,JN,KN
 510 I8=3*ME
 520 IJ(I8-2)=IN
 530 IJ(I8-1)=JN
 540 IJ(I8)=KN
 550 IFABS(JN-IN)>MXTHENMX=ABS(JN-IN)
 560 IFABS(KN-JN)>MXTHENMX=ABS(KN-JN)
 570 IFABS(KN-IN)>MXTHENMX=ABS(KN-IN)
 580 NEXTME
 590 NW=(MX+1)
 600 NT=NW+NN
 610 DIMA(NT,NW),Q(NT),CV(NT),X(NN),Y(NN)
 620 PRINT"FEED NODAL COORDINATES":PRINT
 630 FORII=1TONN
 640 PRINT"FEED X COORD FOR NODE";II
 650 INPUTX(II)
 660 PRINT"FEED Y COORD FOR NODE";II
 670 INPUTY(II)
 680 NEXTII
 690 IFNF=0THEN790
 700 PRINT:PRINT"FEED IN THE BOUNDARY CONDITIONS":PRINT
 710 FORIJ=1TONF
 720 PRINT"NODAL POSITION";IJ
 730 INPUTN9
 740 PRINT"VALUE OF THE FUNCTION AT THIS NODAL POSITION"
 750 INPUTPH
 760 NS(IJ)=N9
 770 NQ(IJ)=PH
 780 NEXTIJ
 790 FORME=1TOMS
 800 PRINT:PRINT"ELEMENT NO.";ME;"UNDER COMPUTATION":PRINT
 810 I8=3*ME
```

```
820 IN=IJ(I8-2)
830 JN=IJ(I8-1)
840 KN=IJ(I8)
850 B(1)=Y(JN)-Y(KN)
860 B(2)=Y(KN)-Y(IN)
870 B(3)=Y(IN)-Y(JN)
880 C(1)=X(KN)-X(JN)
890 C(2)=X(IN)-X(KN)
900 C(3)=X(JN)-X(IN)
910 DL=X(IN)*B(1)+X(JN)*B(2)+X(KN)*B(3)
920 DL=ABS(DL)
930 IFQ7=1THEN960
940 PRINT:PRINT"FEED THE VALUE OF (Q) FOR ELEMENT";ME
950 INPUTQ
960 FORI=1TO3
970 FORJ=1TO3
980 SK(I,J)=0.5*C8*(B(I)*B(J)+C(I)*C(J))/DL
990 NEXTJ
1000 SQ(I)=0.5*Q*DL/3
1010 NEXTI
1020 I1=IN
1030 J1=JN
1040 K1=KN
1050 FORJJ=1TO3
1060 IFJJ=1THENNR=I1
1070 IFJJ=2THENNR=J1
1080 IFJJ=3THENNR=K1
1090 FORKK=1TO3
1100 IFKK=1THENN9=I1
1110 IFKK=2THENN9=J1
1120 IFKK=3THENN9=K1
1130 NK=N9+1-NR
1140 IFNK<=0THEN1160
1150 A(NR,NK)=A(NR,NK)+SK(JJ,KK)
1160 NEXTKK
1170 Q(NR)=Q(NR)+SQ(JJ)
1180 NEXTJJ
1190 IFXH=0ANDXQ=0THEN1250
1200 PRINT:PRINT"HAS ELEMENT WITH NODES";IN;"-";JN;;"-";KN;" A BOUNDARY WHERE E
ITHER";
1210 PRINT" HL OR QG HAS A VALUE? IF YES, TYPE Y,ELSE TYPE ANY OTHER KEY▮"
1220 GETB$:IFB$=""THEN1220
1230 IFB$<>"Y"THEN1250
1240 GOSUB1750
1250 NEXTME
1260 IFNF=0THEN1320
1270 FORII=1TONF
1280 N9=NS(II)
1290 Q(N9)=1.0E12*NQ(II)*A(N9,1)
1300 A(N9,1)=A(N9,1)*1.0E12
1310 NEXTII
1320 FORII=1TONN
1330 CV(II)=Q(II)
1340 NEXTII
1350 PRINT:PRINT"THE SIMULTANEOUS EQUATIONS ARE NOW BEING SOLVED":PRINT
1360 GOSUB1500
1370 PRINT:PRINT"THE NODAL VALUES OF THE FUNCTION ARE AS FOLLOWS:":PRINT
1380 FORI=1TONN
1390 UG(I)=CV(I)
1400 PRINT:PRINT"VALUE AT NODE";I;"=";UG(I)
1410 PRINT:PRINT"TO CONTINUE, PRESS ANY KEY":PRINT
1420 GETA$:IFA$=""THEN1420
1430 NEXTI
1440 PRINT:PRINT"DO YOU REQUIRE SLOPES OF THE FUCTIONS AT THE NODES FOR EACH EL
EMENT?";
1450 PRINT" IF YES, PRESS Y, ELSE PRESS ANY OTHER KEY":PRINT
1460 GETC$:IFC$=""THEN1460
1470 IFC$<>"Y"THEN1490
1480 GOSUB2220
1490 END
1500 FORII=1TONN
1510 IK=II
1520 FORJJ=2TONW
1530 IK=IK+1
1540 KN=A(II,JJ)/A(II,1)
```

```
1550 JK=0
1560 FORKK=JJTONW
1570 JK=JK+1
1580 A(IK,JK)=A(IK,JK)-KN*A(II,KK)
1590 NEXTKK
1600 A(II,JJ)=KN
1610 CV(IK)=CV(IK)-KN*CV(II)
1620 NEXTJJ
1630 CV(II)=CV(II)/A(II,1)
1640 NEXTII
1650 FORIZ=2TONN
1660 II=NN-IZ+1
1670 FORKK=2TONW
1680 JJ=II+KK-1
1690 CV(II)=CV(II)-A(II,KK)*CV(JJ)
1700 NEXTKK,IZ
1710 RETURN
1750 IFXH=0THEN2030
1760 PRINT:PRINT"ON HOW MANY BOUNDARIES OF THIS ELEMENT HAS HL A VALUE?":PRINT
1770 INPUTLH
1780 IFLH=0THEN2030
1790 IFLH=1ORLH=2ORLH=3THEN1820
1800 PRINT:PRINT"LH MUST BE 1 OR 2 OR 3":PRINT
1810 GOTO1760
1820 PRINT:PRINT"INPUT THE NODES ON DEFINING EACH BOUNDARY ON WHICH HL HAS A VA
LUE";
1830 PRINT", TOGETHER WITH THE VALUE OF HL & T(INFINITY)":PRINT
1840 FORII=1TOLH
1850 PRINT:PRINT"NODES DEFINING BOUNDARY";II
1860 INPUTIL,JL
1870 PRINT:PRINT"FEED VALUE OF HL":PRINT
1880 INPUTHL
1890 PRINT:PRINT"FEED VALUE OF T(INFINITY)":PRINT
1900 INPUTIT
1910 XL=SQR((X(IL)-X(JL))↑2+(Y(IL)-Y(JL))↑2)
1920 Q(IL)=Q(IL)+XL*HL*IT/2
1930 Q(JL)=Q(JL)+XL*HL*IT/2
1940 A(IL,1)=A(IL,1)+XL*HL/3
1950 A(JL,1)=A(JL,1)+XL*HL/3
1960 NK=JL-IL+1
1970 IFNK<=1THEN2000
1980 A(IL,NK)=A(IL,NK)+XL*HL/6
1990 GOTO2020
2000 NK=IL-JL+1
2010 A(JL,NK)=A(JL,NK)+XL*HL/6
2020 NEXTII
2030 IFXQ=0THEN2210
2040 PRINT:PRINT"ON HOW MANY BOUNDARIES OF THIS ELEMENT HAS QG A VALUE?:?
2050 INPUTGQ
2060 IFGQ=0THEN2210
2070 IFGQ=1ORGQ=2ORGQ=3THEN2100
2080 PRINT:PRINT"GQ MUST BE 1 OR 2 OR 3":PRINT
2090 GOTO2040
2100 PRINT:PRINT"INPUT THE NODES DEFINING EACH BOUNDARY OF THIS ELEMENT ON WHIC
H";
2110 PRINT" QG HAS A VALUE, TOGETHER WITH THE VALUE OF QG":PRINT
2120 FORII=1TOGQ
2130 PRINT:PRINT"NODES DEFINING BOUNDARY";II
2140 INPUTIL,JL
2150 PRINT:PRINT"FEED VALUE OF QG":PRINT
2160 INPUT QG
2170 XL=SQR((X(IL)-X(JL))↑2+(Y(IL)-Y(JL))↑2)
2180 Q(IL)=Q(IL)-XL*QG/2
2190 Q(JL)=Q(JL)-XL*QG/2
2200 NEXTII
2210 RETURN
2220 FORME=1TOMS
2230 I8=3*ME
2240 IN=IJ(I8-2)
2250 JN=IJ(I8-1)
2260 KN=IJ(I8)
2270 B(1)=Y(JN)-Y(KN)
2280 B(2)=Y(KN)-Y(IN)
2290 B(3)=Y(IN)-Y(JN)
2300 C(1)=X(KN)-X(JN)
```

```
 2310 C(2)=X(IN)-X(KN)
 2320 C(3)=X(JN)-X(IN)
 2325 DL=ABS(X(IN)*B(1)+X(JN)*B(2)+X(KN)*B(3))
 2330 UL(1)=UG(IN)
 2340 UL(2)=UG(JN)
 2350 UL(3)=UG(KN)
 2360 FORII=1TO3
 2370 BM(1,II)=B(II)/DL
 2380 BM(2,II)=C(II)/DL
 2390 NEXTII
 2400 FORII=1TO2
 2410 SL(II)=0
 2420 FORJJ=1TO3
 2430 SL(II)=SL(II)+BM(II,JJ)*UL(JJ)
 2440 NEXTJJ
 2450 NEXTII
 2460 PRINT:PRINT"THE SLOPES WITH RESPECT TO THE X AND Y AXES, FOR ELEMENT ";
 2470 PRINTIN;"-";JN;"-";KN;" ARE AS FOLLOWS:":PRINT
 2480 PRINTSL(1);SL(2)
 2490 PRINT:PRINT"TO CONTINUE, PRESS ANY KEY":PRINT
 2500 GETA$:IFA$=""THEN2500
 2510 NEXTME
 2520 RETURN
READY.
```

APPENDIX 15 Program Running

The programs should be loaded by the methods described in the appropriate manufacturer's manuals.
The program codenames for the programs described in Chapters 1 to 14, and listed in Appendices 1 to 14 are as follows:

Chapter and Appendix	Program Codename
1	TRUSS
2	CONTINUOUS BEAMS
3	PLANEFRAME
4	SPACETRUSS
5	VIBPTR
6	VIBCB
7	VIBRGPF
8	VIBSPTR
9	3D-RIGFRAME
10	GRILLAGE
11	BULKHEAD
12	THINCONE
13	PLANESTRESS&STRA
14	LAPLACE

The output for a typical problem, (Example 1.1), is given below.

```
THE NODAL DISPLTS ARE

-1.72385763E-12 -1.72385763E-12

-.957627776 -5.64203552

3.06091934E-12 -3.69780938E-12

-1.95762778 -16.0997634

FOR SUCCESSIVE VALUES OF NODAL NUMBERS OF

MEMBERS AND FORCES, PRESS ANY KEY

1  2 -3.29983165
2  4 -1
2  3 -1.49071198
3  4 -7.07106781
```

Index